PERFORMANCE EXCELLENCE

in Manufacturing and Service Organizations

Proceedings of the
Third Annual Management Accounting Symposium
San Diego, California
March 1989

AMERICAN ACCOUNTING ASSOCIATION
5717 Bessie Drive
Sarasota, Florida 34233

National Association of Accountants   American Accounting Association
Committee on Academic Relations   Management Accounting Section

PERFORMANCE EXCELLENCE

in Manufacturing and Service Organizations

Proceedings of the
Third Annual Management Accounting Symposium
San Diego, CA
March 1989

Edited by Peter B. B. Turney

TABLE OF CONTENTS

PREFACE

On March 10-11, 1989, a symposium entitled "Performance Excellence in Manufacturing and Service Organizations" was held in San Diego, California. Eleven papers were presented to an audience of more than one hundred and fifty business executives, consultants and academics. The presentations were delivered by fourteen individuals, nine from academia and five from consulting or industry. Nine of the papers are included in these proceedings.

The symposium was the third in a series of symposia co-sponsored by the Management Accounting Section of the American Accounting Association and the Committee on Academic Relations of the National Association of Accountants. The first symposium, which occurred in 1987, focused on the failure of management accounting to adapt to today's competitive environment. The second symposium, delivered in 1988, exposed the academic membership of the American Accounting Association to new teaching materials in the area of product costing. The third symposium was an opportunity to move beyond the failures of management accounting, and to focus on new tools that are available to meet today's competitive needs. The overall theme of the third symposium was therefore decidedly up-beat.

Judging by the number of questions and the enthusiasm of the audience, the symposium was well received. The program benefited from the wide variety of topics presented, and the diversity of backgrounds of the speakers and the audience. The presentations from the world of operations management showed how closely linked their research is with ours. It demonstrated how much we can benefit from links with this field.

Much of what was discussed in the symposium differed in terminology or scope from traditional management accounting. Tools such as Quality Function Deployment (Quevedo), Activity-Based Costing (Cooper and Turney), Activity Analysis (O'Brien, Johnson, Fults, Jackson), Statistical Process Control (Reeve) and Non-Financial Performance Measurement (Schonberger, Hall, Vollman) have been given little or no emphasis in curriculum or research in the past. They are likely, however, to be the centerpiece of management accounting for the future.

A common theme for many of the papers was the need to satisfy the customer. An important test of the effectiveness of any organization is its ability to generate customer value (Johnson). Each activity must be judged for its contribution to the needs of the customer. This is likely to be a recurring theme in management accounting in the nineties.

The papers show that manufacturing and service organizations face common problems and benefit from the same tools. What do their products cost? How can they become more competitive? How do they focus their activities on satisfying the needs of their customers? These problems are common to manufacturing and service organizations alike. Tools such as Activity-Based Costing and Quality Function Deployment work just as well in a service organization as in a manufacturing company. The Activity Analysis in a telecommunications company described in the Johnson, Fults and Jackson paper parallels a similar exercise in a manufacturing organization described in O'Brien's paper.

The symposium ended with a panel of speakers fielding questions from the audience. The types of questions and the contrasting answers given by the panelists suggests there are many issues still to be resolved. The role of cost in achieving performance excellence, for example, was a major issue. Is cost still the focal point of management accounting, or should it play a reduced role as suggested by some of the speakers?

I am grateful for the contributions of the fourteen individuals who prepared papers and delivered presentations. These individuals represented my first choice of participants for the symposium. Not one person turned down my invitation to be on the program. The willingness of the speakers to contribute and share their knowledge and time with us is much appreciated. It certainly made the success of the program possible.

A number of individuals helped in the planning and execution of the symposium. H. Thomas Johnson of Portland State University, helped identify speakers for the program. Shane Moriarity of the University of Oklahoma, Chairman of the Management Accounting Section of the American Accounting Association, assisted in the planning of the program. Ralph Estes of Wichita State University, Chairman of the Research Committee of the Management Accounting Section, and Hadassah Slominsky of the National Association of Accountants provided valuable support. Van Ballew of San Diego State University organized the Book Fair. Paul Gerhardt of the American Accounting Association arranged for the hotel. Linda Sydenstricker of the American Accounting Association handled the advanced registrations, arrangements and the registration desk. Anne Riley of Portland State University and Pat Calomeris of the American Accounting Association helped in the preparation of the proceedings. I thank all of them for their valuable contributions.

The symposium was co-sponsored by the National Association of Accountants. I gratefully acknowledge their generous financial support. The close relationship between the Management Accounting Section of the American Accounting Association and the National Association of Accountants is an important source of strength for the Section. The success of this symposium owes much to this relationship.

Portland, Oregon   Peter B. B. Turney
February, 1990

PERFORMANCE EXCELLENCE
in Manufacturing and Service Organizations

WORLD-CLASS PERFORMANCE MANAGEMENT

Richard J. Schonberger
Schonberger & Associates, Inc.
Seattle, Washington

ABSTRACT

Led by manufacturing, world-class companies are remaking their cost control and product costing systems. The conventional transaction-driven system is costly to administer, fails to control cost, and usually yields erroneous product cost data.

The new system controls causes of cost directly, where the causes occur; it is non-monetary, largely visual, and operator centered. It stimulates continual improvement and thus should be implemented widely and quickly.

While the control system needs to operate all the time, product costs are needed infrequently—for bidding, pricing, or go/no-go decisions. The new thinking is to just conduct a quick cost audit when a product cost is needed; abandon cost transactions, except to satisfy minimal legal and regulatory requirements. Audits emphasize assigning overhead to products accurately, which generally means assigning much more per unit to low volume than high volume products.

The world-class performer—person, group, or company—must improve at a rapid rate, and keep it up. Virtually no Western company has done that in recent decades. Not only did we not know *how* to sustain a high improvement rate, we didn't even know *what* to improve. Cost problems, market-share losses, and migration of whole industries to the Far East were evident; root causes, poor quality being first on the list, were not.

With elevated costs and things not being done right (high error rates, delay-prone design and production, equipment in sad shape, employees ill-trained, information inaccurate), industry's response was to add controllers—layer upon layer of them. The bloated staff groups in our corporations, large and small, were not trying for big wins so much as holding things together. Their solutions were reports. And more people to review the reports. And still more to devise still more reports. Industry was in dire need of a fresh breath of sanity, and many people knew it.

Sanity has arrived. Enlightened firms chopped layers out of their organization structures and, in tandem, rolled out employee involvement (EI) campaigns. It was involvement in process control and improvement, not mere participation in vacation scheduling and what color to paint the walls. For EI to work, employers had to increase training budgets, from essentially zero for shop-floor people to a few percent of the payroll.

An outpouring of training materials—new ideas for control and improvement—made it rational to spend those sums on training. Line employees were schooled in the new techniques at the same time as, or right after, their colleagues in management and staff positions. They all learned about measuring process output, plotting data, brainstorming, process capability, next-process-as-customer, quick setup, cellular organization, kanban, and dozens more. Companies that have gone through this are in control and improving smartly. The old control system, based on cost-watching, has given way to direct controls on the causes of cost.

OPERATING CONTROLS

The new direct controls are simple, on-the-spot, and mostly visual; no more long, winding information trails from operations to analysts (and their computers) to layers of managers to supervisors and finally back to the source. The visuals, all large, easy to see, prominently placed in the work zones, and run mainly by operators, include:

- Performance graphs—on throughput time, idle inventory, flow distance, yield, rework, etc.
- Statistical process control charts
- Slow and stop actuators and lights—and charts to record causes of slow-ups and stoppages
- Pareto and fishbone charts (and other analysis sheets) for on-going improvement projects
- Marked off zones showing exactly where everything—tools, containers, parts, fixtures, etc.—is to be placed
- Color-coding for easy identification (e.g., fittings for lubrication, look-alike parts, tools)
- Maintenance check sheets and diagrams
- Scheduling displays, including the visual kanban scheduling system
- Cross-training displays
- Procedural diagrams and templates tacked to walls and work benches
- Photos of people winning awards

These controls, collectively, aim at preventing costly mix-ups, defects, rework, scrap, delays, queues, storages, multiple handling, and searches. They aim, as well, at recording all disturbances and fixing the causes. Finally, they recognize and celebrate progress, improvement, and contributions. What more could we ask of a control system?

These world-class operational controls are non-monetary: a comprehensive non-cost system that controls cost—and brings in more revenue from customers attracted by better quality, quicker response, greater flexibility, and lower cost goods and services. The cost system had some of these goals, too, but the cost system was too delayed, too aggregated, and too far removed from the action to be effective.

In remaking our performance management system, operational controls are just one aspect—but one to be implemented right away since it stimulates continual improvement. Another aspect is capital equipment, a resource that has been managed based on unsound concepts of planning and measurement.

CAPITAL EQUIPMENT

World-class equipment policies go rather opposite to trends in the last few decades. Instead of fewer, faster, and larger pieces of equipment, it is more, slower, and smaller—the *economy of multiples* concept. A few of the reasons are as follows:

- More machines. Much of industry (factories at first, but now offices, too) is being reorganized into work cells. In the cellular organization, each cell is dedicated to a narrow family of products with its own production team, and each must have its own personal computer, bonding oven, welder, or wave-solderer. A huge main frame, massive oven, super-welder, or super-wave can only be in one place at a time—can't fit into multiple cells or be product-focused.

• Slower machines. Dedicated machines and cells may each be run at a speed close to the rate of use of the next process—which ultimately is the sales rate to final customers. The supermachines favored in the recent past often pump out units (only *one* product model at a time) at rates far in excess of use rates— sometimes 10,000 times greater. That just means putting the production into costly storage—and hoping the forecast is not too far off so it can be sold some day. The concept of running machines at the rate of use at the next process instead of full blast signals the death of a universal tool of the old control system: the machine utilization report.

• Small machines. Continual improvement, quick introduction of new products, and quick-change response to demand recommends machines that can be moved easily and often. Some plant managers, imbued with the new thinking, are putting machines on heavy-duty casters, rather than bolting machines to the floor. Of course, a machine must be as large as necessary to accomplish its required task. Many kinds of machines have grown far larger: ovens, dye vats, sterilizers, retorts, mixing tanks, and cookers that hold thousands of units of product; sheet and roll processing machines eight feet wide, which send product to successive slit and chop stages to get it down, finally, to saleable sizes; all-in-one fabrication machines and testing equipment so massive as to require special flooring. These kinds of equipment are beginning to look like dinosaurs.

By itself, more machines (for multiple cells) would mean more equipment investment. The combination, more but slower and smaller, does *not* mean more equipment investment. The cost of five small, simple machines may or may not be greater than one supermachine with the same capacity. Regardless of that, the five are likely to greatly cut the total cost when disbursed into five cells. Since each cell is dedicated, there is little lost time or labor for set-ups and changeovers, little inventory and handling, and little of the many related overhead costs.

Machines in cells also may cut the number of operators, since one person usually can tend several unlike machnes in a round-robin route; further, machines in cells can easily be modified for automatic unloading, transfer, and loading. (At Allen Bradley in Milwaukee, a few injection molding machines have been moved into cells. In one cell, each molded plastic part is blown by a whiff of air through a tube to the nearby assembly station, next process in the cell.) Perhaps most importantly, when machines are in a cell, operators can often detect defects immediately—bad part from machine A can't be processed in machine B; and there are good chances of fixing problems while evidence of causes is fresh. Quality control is direct, fast, and largely in the hands of the operators.

And what about the supermachine already owned? Make it run perfectly, preventively maintain it so it never breaks down, learn to set it up quickly, plot its output on statistical quality control charts, assign it to make only a few high volume models, slow it down, run it intermittently. *Don't* try to maximize its utilization; that may lower one unit cost (depreciation), but it will raise at least a dozen more.

PRODUCT COSTING

Non-cost controls on cost do not quite eliminate managerial needs for a cost system. There can be tactical value in finding out, once in a while, what each product or product type costs to produce. Uses include pricing, bidding, and making go/

no-go decisions. These kinds of decisions are infrequent, perhaps once or twice a year. Therefore, the world-class company has no need to charge labor, materials, and overhead to products made as they are made, day in and day out.

Today's product costing system is simply a cost audit, when needed. In a medium-sized company, that eliminates thousands of cost transactions (mainly labor and materials) per month. Furthermore, a quick cost audit is likely to provide far more accurate product costs than transaction-driven cost systems ever did.

The reasons why the conventional cost system is inaccurate are by now well known (if only to those in industry and academia who are up-to-date). It aggregates and averages and sprinkles overhead upon products sort of like falling raindrops, rather than based on how products actually draw upon overhead resources. The system is acceptable for legal and regulatory purposes but can and often has had disastrous consequences when used for tactical decisions. An oft-cited example is Western industry's give-away to the Far East of whole high volume industries, such as VCRs and memory chips. Companies that abandoned those products did so partly because their quality was not up to snuff and their overall cost structure was too high (not so much the case today).

But why did they get out of high volume products and retain low volume, high variety product lines? Mainly because the conventional cost system lied. It grossly misallocated bloated overhead costs, applying far too much overhead to high volume products—which really don't require much overhead per unit at all. High volume products zip through the processes with relative ease, whereas each low-volume order is a special case that has to be mother-henned.

The old cost and control system does not disappear overnight. Even in the most enlightened firms—those with a full set of direct visual controls—the conventional system resists quick extinction. What happens in the transitional years? The old, still-standing system becomes irrelevant. Like sort of a reverse Gresham's Law, the good drives out the bad, because the good controls—those that control *causes*—have so many vigorous vocal champions. Still, the conventional approach should not be allowed to linger. It costs too much to administer.

STATE OF IMPLEMENTATION

The new approaches to control and costing are only a few years old. Earliest developments, circa 1984, occurred largely in U.S. companies like Hewlett Packard (certain divisions), Zytec (a specialty power supply manufacturer in Minnesota), and Harley-Davidson. People like myself played small early roles, offering an idea or two and spreading the word via seminars and consultancy. Before long, a few management accounting academics were in the thick of it—especially Thomas Johnson, now at Portland State University, and Robert Kaplan and Robin Cooper of Harvard University.

Even though the new approaches call for radical changes, there is surprisingly little squabbling and disagreement about the nature of the new system (among those who have got over their initial shock). This parallels other related developments, especially the worldwide enthusiasm for total quality, just in time, cellular organization, quick changeover, cross training, employee involvement, and so forth.

In fact, some of the simplifications in accounting depend on prior improvements in operations. For example, a few dozen Western plants have been "allowed" (by their accountants and outside auditors) to halt "WIP (work in process) tracking." These plants had cut their production lead times from weeks to days, which made front-door/back-door inventory accounting good enough; that is, accurate enough inventory valuation to meet regulatory requirements.

Reports that Japanese industry uses different accounting systems are, I think, mostly inaccurate. Japan's accounting systems are quite like Western ones. The same accounting system requires far fewer transactions in top Japanese companies because of short lead times and years of attention to making things work. Potent visual controls on factory floors render periodic reports largely irrelevant, but they are still churned out.

While only a few companies have extensively implemented the new control/costing system, it seems safe to say that at least initial planning is taking place in most of the Fortune 500 industrial firms in the United States. European seminars on the subject sprang up in late 1988 and early 1989, and now, quickly, it's become a worldwide movement. Country-specific accounting conventions and practices make no difference, because this is a movement toward reducing control and costing to simple basics, and basics are basics everywhere.

PROFESSORS, CUSTOMERS, AND VALUE: BRINGING A GLOBAL PERSPECTIVE TO MANAGEMENT ACCOUNTING EDUCATION

H. Thomas Johnson
Retzlaff Professor of Cost Management
Portland State University
Portland, OR 97207-0751

ABSTRACT

This paper explores the divergence between the traditional methods taught in current management accounting education and the new activity management philosophy being explored in the business community today.

Traditional management accounting places primary emphasis on production output and cost. The traditional economic model encouraging maximum utilization of resources leads firms to minimize unit production costs by increasing volume and thus increasing total production costs. Traditional manufacturing systems use financial planning systems for decision making purposes and push financial goals of the company downward to the production level. Thus operational decisions must pass a test of financial not operational soundness before being implemented. Traditional systems also encourage decision making in a static environment of constrained resources. Such a view prevents seeing potential changes a company can make by changing its operational parameters.

Activity management centers around maximizing customer value as quickly and efficiently as possible. By examining operational activities, eliminating those that do not contribute to stated customer demands, and striving to continuously add value as quickly as possible, total costs are minimized and directed to the organization's primary purpose. Efficiency of operations are tracked with charts and reinforcement, and production teams are empowered to make changes in process specifications and activities. Embracing continuous change at the operational level will enhance the opportunity for global success.

Recent doubts about the ability of traditional management accounting to provide information businesses need to compete in the global economy ought to draw management accountants in business and in universities closer together. One would expect management accounting professors to ask management accountants in business about current problems and proposed solutions. Similarly one would expect business executives to ask professors what their research suggests about solutions to management accounting problems.

Unfortunately such exchanges seldom occur. Businesses rarely turn to professors of management accounting for assistance, and university management accounting programs seldom address the issues vexing business executives. This relative

I want to thank Peter B. B. Turney for comments on earlier drafts of this paper without implicating him in errors, omissions, and inconsistencies that remain.

absence of communication between universities and business exists for one major reason. Most academic teaching and research in management accounting concentrate on issues raised by economists and psychologists, not on the urgent concerns of today's business people. Such issues, while fascinating to other academicians, seldom encourage professors to address the needs of modern management accounting in business. To effectively address these needs we must bridge the gulf between universities and businesses.

In an attempt to bridge that gulf, this paper will explain why traditional management accounting practices no longer work in modern businesses. Having pointed out management accounting's failures the paper will suggest how universities might adjust their programs to address these deficiencies. Presumably if academicians are cognizant of problems in a business organization they will strive to identify solutions to these problems.

GLOBAL COMPETITION AND MANAGEMENT ACCOUNTING

Most of the deficiencies in traditional management accounting emanate from two developments, both of which took place after 1970. First, new information and communication technology enabled customers (both final goods consumers and industrial buyers of intermediate inputs) to find and get access to the *best* of what they wanted wherever it was available in the world. No longer could suppliers survive simply by doing as well as local competitors. Now suppliers had to compete globally. Second, new production techniques—largely from Japan—enabled suppliers to profitably sell goods of unprecedented quality at low, competitive prices. These new production techniques, focused as they were on satisfying customer wants, sharply increased the terms of competition just at the time customers were gaining access to global markets. The impact of both developments—production techniques and communication and information technology—hit American manufacturers with devastating force in the late 1970s and early 1980s.

Perhaps management accounting information should have allowed manufacturers to anticipate the changes occurring after 1970. Certainly it should have given manufacturers some indication of how to respond to these two major developments once they had taken place. That management accounting information did neither evidences its inability to help companies compete in a global economy.

Traditional management accounting failed American manufacturers in the 1970s and 1980s largely because it did not respond to the information needs of a new competitive environment. Traditional cost accounting and budgeting tools define competitiveness as "beating your competitors on cost." According to that perspective, you control unit costs and achieve profitability by producing all you can sell at a price that exceeds variable cost. What you sell is designed and built primarily with an eye to products, processes, and costs of your competitors. Scant attention is paid to customer perceptions of value. Before 1970 most American businesses considered the customer merely as someone you persuade to buy your output.

After 1970 a new competitive environment gradually appeared that required information not provided in traditional management accounting systems. Competitiveness in that environment requires two steps: first, listening to the voice of the customer, and second, continuously eliminating waste (i.e., work that adds no value to the customer). Satisfying the customer was manifest in the late 1970s and early 1980s by a worldwide concern with quality, and in more recent years by a concern to be flexible. Continuously eliminating waste is associated with programs for total quality control, just-in-time production, and employee involvement. Invariably these two steps are interrelated: satisfying the customer is achieved by eliminating waste (and vice versa); quality and flexibility are achieved by just-in-time programs that empower workers and reduce nonvalue work; just-in-time goals are achieved by efforts to increase flexibility; and so forth.

Traditional management accounting systems provide virtually no information about value received by the customer or about value added in the workplace. They record revenue, but that simply reflects what the customer paid for a product. Revenue gives no indication if the product satisfied the customer's needs, nor if the customer might have preferred something else had it been available. Traditional management accountants also record costs, but accounting costs only give information about the object of expenditure, not about the work that causes cost. Traditional cost information does not indicate if work adds value to a customer, nor does it tell a would-be global competitor how to continuously improve at adding value to the customer.

Lulled into complacency by traditional management accounting signals, American manufacturers were surprised by the high-quality and low-price of Japanese goods that began to flood our markets in the late 1970s. The Japanese couldn't sell those products at their prices and still make money, according to cost information based on best-practice American operations.

That is where we missed the point. The Japanese weren't using American operating procedures. By the late 1970s their best-practice exporters were doing things quite differently than we had ever imagined. We did not really understand what they were doing until the early-to-mid-80s. One thing that prevented us from seeing what they were doing was the competitive philosophy imbedded in our traditional management accounting systems.

Our management information reinforced the philosophy of controlling unit costs by producing more and faster. Their competitive philosophy, apparently not encumbered by American-style management accounting information, emphasized producing only what was demanded, always on schedule, and without error or waste. They produced excellent products at low cost. We, on the other hand, failed to see that our efforts to control unit cost hampered product quality and, by causing chronic overhead creep, raised total costs.

We are beginning to understand many differences between these two management philosophies, although it has taken time and we still have far to go. We can list many features usually found in best-practice Japanese operations but seldom in ours: production in small lots, fast changeovers, processes laid out according to flow of work rather than type of work, and responsibility for decisions delegated to the lowest level. And we know how outcomes of their best-practice operations usually differ from ours: less space, less time, less work in process, less supervision, less accounting, less scheduling, less cost, higher quality, better adherence to schedules, and more enjoyment of work at all levels.

How do well-run Japanese operations achieve these features and outcomes? We have learned enough in the past decade to know their accomplishments are not just a by-product of unique cultural traits. They succeed by learning what will satisfy customers, identifying what work is needed to provide that satisfaction, and finding (and eliminating) unnecessary work. It seems they achieve superior outcomes by assiduously managing the value that work provides customers—something many authorities believe we can do just as well if we put our minds to it.

Many people who have studied what Japanese say about managing work have suggested that American managers might achieve the same goals by pursuing what is called "activity management." The key is to manage activities—the work people do that provides value and consumes resources. The idea is to identify the activities needed to satisfy customer wants and to compare that flow of activities with the activities presently performed in an organization. The objective is to find and eliminate unnecessary activities now being performed and to do more of what adds value. American companies that conduct activity audits generally find 50 to 90 percent of the work they do adds no value to customers, although it causes a great deal of cost.

Discovering a high percentage of nonvalue work is one shocking revelation of an activity audit. Greater shock hits when companies realize that years spent managing costs never prompted them to ask if the work causing these costs had any value for customers. It seems incredible that so much nonvalue work can occur in companies that have assiduously managed costs for long periods, in some cases for decades. But the reason is right under our noses. *Most nonvalue work is caused by the way companies control unit costs of output in the many activities it takes to design, make, and sell a product.*

American companies attempt to control unit costs by specializing and subdividing work according to processes and then increasing the output and the scale of every process they conduct—e.g., mixing, grinding, mail sorting, welding, heat-treating, parts ordering, invoice preparation, assembly, and so forth. Underlying most of our operational budget and cost control systems is the idea, perhaps first enunciated by Adam Smith, that large scale and high utilization reduces costs of direct labor, machine-time, and capital outlays per unit of output. However, this idea ignores the adverse impact of large scale and high utilization on inventories, lead times, space utilized, distances things are moved, schedule delays, mistakes, rework, and customer complaints. That adverse impact is reflected in overhead costs. Moreover, those overhead costs increase as our effort to control unit costs by producing more and faster increases the amount produced in excess of current demand. Indeed, the majority of indirect cost in American companies is caused by work that has no value to the customer, and much of that work entails producing unwanted output.

To be world-class competitors, companies must recognize the earmarks of a globally competitive enterprise and they must know how to achieve those earmarks. The chief characteristic of a globally competitive operation is *continuous flow of value-adding activity as fast as possible.* To achieve that state, according to Taiichi Ohno, "father" of the Toyota production system, you must do two things: *produce only what customers want, when they want it* and *be able to stop any process to correct mistakes.*

Activity management is a viable strategy for achieving the status of global competitor. By adroitly managing activities a company produces at low cost only as much as is needed for current demand. When demand is met, production stops and work shifts to training, practice, and improvement—improvement in understanding what the customer wants and in understanding how to meet those wants. In that world there are no fixed costs and no product life-cycles, only continuous regeneration of opportunities to profitably employ managed resources. Activity management—the pathway to that world—lifts companies from the stagnant plateaus reached after they have cut lot sizes to cut inventory and then stopped; after they have focused their product lines to reduce complexity and then stopped; after they have slashed budgets to cut costs and then stopped; after they have achieved goals set by optimization targets and then stopped. When will a company know it has reached the world of global competitiveness? When it thinks of low cost in terms of continuous and rapid flow of value-adding work, not in terms of cutting unit costs.

To become world-class competitors, companies must understand how activities contribute value to the customer. Understanding the cause-and-effect relations between activities and value requires either instinct or information (or both). Western management accountants who have written about activity management tend to stress information. This emphasis on information is important, but I believe it is just a first step on the road to understanding what it takes to be competitive in the global economy. Japanese authorities, while not ignoring the role of information, seem to stress the importance of instinct. We have a long way to go, I believe, before we truly understand this instinct and can replicate (or improve) their practices.

Meanwhile, many American management accountants, especially those in universities, are not even familiar with what Western authorities say about the role of information in activity management. Let's summarize what these authorities say by defining and comparing today's three most popular approaches to the subject:

1. Activity management (AM)—measure nonfinancial outcomes of activities;
2. Activity cost accounting (ACA)—measure cost of activities;
3. Activity-based product costing (ABPC)—measure cost of products through activities.

All three approaches start with tracking the chain of activities that adds value to customers (although they do not all define "activity" the same way). The first two approaches, AM and ACA, focus on finding waste in activities (i.e., nonvalue work), as that concept is articulated by Japanese authorities—especially those associated with designing the "Toyota production system." They differ, however, in the way they define waste in activities. The AM approach sees waste to some degree in all human work at all times,[1] whereas the ACA approach defines certain types of activities as inherently wasteful (e.g., moving, storing, waiting, reworking, etc.). The ABPC approach focuses on activities as the logical place to assign costs in an organization (since activities consume the resources that cause costs).[2] Its primary objective is to overcome distortions found in traditional accounting-based product costs. It reduces these distortions by basing the costs of products on the costs of activities it takes to design, make and deliver a product.

The ACA and ABPC approaches are unlike traditional management accounting practice in that they compile cost information according to activity. This difference discourages users of ACA or ABPC information from mindlessly linking cost variation to variations in output volume. These two cost-oriented approaches to activity information encourage managers to think of costs in terms of human actions and other "drivers" that cause costs.

The AM approach to managing activities virtually ignores costs. It points attention to waste itself, not to the cost of waste activity. It implicitly assumes that eliminating waste from activities will reduce costs. However, it can make that assumption only because it pursues the strategy of eliminating causes of waste in all activities by *transforming the way we do work*—following precepts of TQC, JIT and so forth. In contrast, the ACA and ABPC approaches, like traditional management accounting practice, do not focus primarily on changing the way a business conducts activities.

---

[1]The AM approach is reflected in the papers in this volume by Hall; Johnson, Fults, and Jackson; O'Brien; Quevedo; Reeve; Schonberger; and Vollmann.

[2]The ABPC approach is reflected in the paper in this volume by Cooper and Turney.

Each of these three approaches, as they are articulated by Western management accountants, provides information about activity *outcomes*—results one hopes to achieve by managing activities. The ACA and ABPC approaches measure cost of activities. The AM approach, not exclusively focused on financial measures, measures outcomes at which Japanese manufacturers seem to excel, such as space occupied, days of inventory, the time it takes to get a job done, total (not unit) cost, and customer reactions. But outcome measures alone don't show global competitors what they must do to make activities continuously more competitive. There is a large gap between measuring outcomes and understanding the practices of competitive world-class producers. That understanding appears more often in writings on the AM approach to activity management than in writings on the ACA or ABPC approaches.

Although costs of activities (ABPC) or costs of activities that are identified with nonvalue work (ACA) may not help businesses identify the actions it takes to be world-class competitors, they are essential information in managing a business. Even if a company could achieve global competitiveness by managing activities without any cost information, situations would still arise when it is necessary to know the cost of something, such as a product, a component, or an inter-departmental service. Reliable cost information of this sort is indispensable to planning, budgeting, all types of "what if" inquiries, and to target costing, a market-based cost concept currently popular for planning and budgeting in Japan. Although a cost system is not needed to estimate total target cost by product, a system is needed to translate that total into costs by component or by department.

Today everyone knows how overhead allocation practices distort the product costs derived from traditional *accounting-based* product costing systems. Activity-based product costing (ABPC) was an important breakthrough that overcame most of those distortions. It is now inconceivable to do planning, budgeting, target costing, and price analysis without ABPC information. Moreover, both ABPC and ACA information have an important role to play in raising consciousness about costs. Like "cost of quality" information, ABPC and ACA information raise consciousness about the cost of waste. But they do no more than cost of quality information does to identify root causes of waste.

The ACA and ABPC approaches do not provide signposts that lead automatically to global competitiveness because they do virtually nothing to alter the competitive perspective that underlies our traditional budget and cost variance control systems. That perspective has managers control costs by specializing and subdividing work and then forcing through as much output as possible in order to achieve scale economies. In a business whose operations are run according to that perspective, most of the work that Japanese managers call nonvalue activity *is absolutely necessary*. In American plants that run according to traditional practices you can not stop all (or even much) moving, storing, inspecting, scheduling, and waiting even though you may say those activities add no value to the customer. Because the way we do things makes those activities necessary, removing (or substantially curtailing) nonvalue activities can not be done without great cost. Thus, *the real challenge facing companies that would compete globally is to understand how to organize operations so as to make nonvalue activity unnecessary*. Once nonvalue work is made unnecessary, then it can be eliminated at little or no cost.

To give an example of what I am talking about, consider an activity such as inspection. Customers derive value from having a product perform according to specifications. Imagine two suppliers, A and B, who can deliver a product that performs equally well but B can do it without inspecting. If all else is equal, supplier B can give customers greater overall value because resources saved from not inspecting will allow them to offer a lower price. In this case inspection is what we call a nonvalue activity.

How will supplier A know what to do to compete with supplier B? If supplier A has done an activity audit or has compiled activity cost information they know that a portion of their manpower and costs are consumed in a nonvalue activity called inspection. Such knowledge can trigger action. If supplier A has information about *activity costs* (ACA), they may take some or all of the following steps: consolidate inspecting in one location (if it is not already) in order to gain economies of specialization and scale; purchase inspection hardware that permits fewer inspectors to check larger volumes of output per day; invest training and software dollars in improved sampling techniques that allow inspectors to check fewer items without loss of reliability; invest in statistical process control technology that allows workers to replace inspectors. If supplier A in fact has *activity-based product cost* (ABPC) information we might expect, in addition to any of the above steps, some or all of the following additional steps: remove or redesign those particular products or processes that consume most of the inspecting activities; focus attention on high volume mature products that require proportionately less inspection work.

The steps taken in the above situations will not necessarily change the conditions that make inspection necessary. They most likely reduce the total or per unit cost of inspection activity. They reduce cost partly by reducing dollars spent on inspection and partly by reducing the demand for inspection work. These are desirable outcomes, but not necessarily enough to make supplier A competitive with supplier B. Assume supplier B had dispensed with inspection by pursuing some or all of the following steps: link processes according to the flow of work it takes to make a product rather than according to the type of work; reduce batch sizes to permit linking of processes; reduce changeover times to make reduction of batch sizes economically feasible; train workers to understand the connection between changeover times, batch sizes, plant layout, and competitiveness. If supplier B is a Japanese company they probably did all this without any cost information, either ACA or ABPC. However, they probably had extensive information on the value to customers of product reliability and information on the need for inspection in their flow of activities.

Some companies that have designed ACA or ABPC systems say that knowing costs by activities is a catalyst that eventually triggers the actions it takes to become competitive. Their traditional management information systems never showed how much it really costs to inspect, to move, to store, and so forth. Charting the flow of activities and estimating the activities' costs enabled them to see the egregious dollar magnitude of waste they had tolerated for so long. This is a salutary discovery, important enough perhaps to warrant investment in activity costing systems. But there is a *risk in trying to make activity cost information serve more than just strategic purposes* such as planning, budgeting, target cost analysis, and pricing studies. Activity cost information as such will do nothing to change the habits of managers who adhere to the traditional competitive perspective. Their instinct will be to reduce unit costs by managing scale and volume of output.

Following traditional tactics to control the costs of activities, managers may attempt to reduce the demand for an activity by seeking economies of scale and throughput in an activity center. For example: to reduce costs of setup activity they may eliminate products sold in low volume and concentrate only on products that can be mass-produced and sold in large volumes;[3] to reduce the cost of making printed circuit boards a company's designers may select a larger-size board in order to facilitate machine-automated as opposed to manual insertion of components;[4] to reduce costs of purchasing, storing, and kitting activities designers may search for opportunities to share components among different products.[5] The list can be extended, but the point is this: reducing unit costs of output by using traditional American management practices to reduce the demand for activities invariably prompts companies to do what is contrary to creating competitive value in the global economy. *These traditional practices prompt managers to economize on activities that essentially add no value to the customer*—e.g., lengthy setups, purchasing, inspecting, and storing. They do not encourage or even prompt managers to think about steps they must take to eliminate such activities, steps such as linking processes, reducing changeover times, and cutting batch sizes. Moreover, activity cost information does not prompt managers to give priority to customer value. In the above example, customers may value small size in the final product that uses a circuit board (e.g., a cellular telephone). The costs saved by designing for automatic insertion might therefore diminish value (and profitability) if the end result is a larger final product.

In the above examples, the steps taken to reduce activity costs may have reduced costs per unit of output in the short run. But in the long run they probably reduced flexibility, decreased quality, and *increased total* costs. This conclusion does not impugn the value of activity cost information for strategic purposes. But it does reinforce my belief that activity cost information does not signal the actions managers must take to insure long-run global competitiveness.

We know very little at the moment about the role information plays in helping world-class Japanese companies achieve favorable outcomes by managing value in the work they do. However, many people who have studied this issue stress the importance Japanese managers place on empowering workers to design and execute the activities they perform. This empowerment underlies the never-ending process that is characterized by the phrase continuous improvement. Continuous improvement appears to be driven primarily by individuals and groups ceaselessly searching for generators of waste in activities. The goal is to remove all work between the customer and any person whose work adds value. Achieving this goal reduces layers in organizational hierarchies, leading eventually, some say, to companies having no "managers"—only people who do value-adding work. This Nirvana-outcome is implied in the idea of every employee being a worker and every worker being a leader.

MANAGEMENT ACCOUNTING EDUCATION IN THE GLOBAL ECONOMY

What should professors do to bring the perspective of global competition into the management accounting curriculum? First they must redefine the phenomena they study. Up to now academics who specialize in management accounting have focused on two phenomena: primarily the "firm" as defined in neoclassical economic theory and also the "individual" as defined by psychologists. Firms and the individual are treated as necessary abstractions from reality—simplified models that permit us to construct theories about reality. The use of these models to study

[3]John Deere (Harvard Business School Case)

[4]Hewlett-Packard: Roseville Networks Division (Harvard Business School Case)

[5]Tektronix: The Portables Instrument Division (Harvard Business School Case)

management accounting seems problematic to me for two reasons: the "firm" as economists define it has little to do with the managed organization we see in the real world; and competitive managed organizations in the real world seem to thrive on the work of teams or groups, not individuals.

In economic theory the "firm" and the "consumer" exist to make tractable a theory of price determination in markets. Given certain assumptions about the profit maximizing and utility maximizing propensities of firms and consumers, economists can derive hypotheses about the behavior of prices in various market settings. The theory of market price determination in its modern mathematical form is an elegant and useful intellectual construct for which several of its creators have justly earned the Nobel Prize. But the primary intent of that theory is not to enhance our understanding of managed organizations. It is intended to enhance our understanding of market behavior. There is a great difference between market and managed direction of economic activity, as the works of Oliver Williamson, Alfred D. Chandler, Jr., and their many followers so eloquently argue. This difference seems to have escaped most management accounting scholars of the past generation.

One important difference, it seems to me, is the role of people in these two modes of economic organization. Market outcomes, in neoclassical economic theory at least, reflect the behavior of individuals. Managed organizations, on the other hand, achieve results through group or team behavior. The power of self-interest seeking individuals to accomplish desirable economic results in competitive markets is awesome, but it is not capable of solving all economic problems all the time. Sometimes we must curb self-interest and work as teams in order to achieve the best economic results. That is the reason for going to the trouble of creating managed organizations, the firm in the real world.

Most economists seem to understand the limited power their "theory of the firm" has to explain the behavior and performance of real-world managed organizations. Even in the subject economists call "industrial organization," where they study the interaction of managed organizations in various market settings, they don't pretend to analyze decision-making inside the organizations. Nevertheless, management accounting academics since the 1950s have made careers out of crossing over this line that economists themselves seem unwilling to cross. They attempt to explain internal decision-making in managed organizations using the profit-maximizing model of "firm" behavior found in the neoclassical theory of market price determination. Chiefly due to the emphasis on scale economies and output volume in that model, the present management accounting curriculum, in my opinion, advocates practices outlined in the previous section that are inimical to competitive behavior in the global economy.

What specific features of university management accounting programs should be changed to adopt the global competitors' perspective? I will restrict my attention in this paper to changes in the content of management accounting courses. In subsequent papers I will outline my recommendations for changing the research agenda pursued by management accountants in academe.

My recommendations for changing the content of management accounting courses all reflect what I see as two fundamental flaws in our present curriculum: 1) emphasis on the output a business produces rather than the activities that consume resources; 2) encouraging managers to control a company's operating activities with accounting information that is designed initially to plan and coordinate financial strategies.

1. *Focusing on the output.* The emphasis on output pervades every topic taught in management accounting courses, beginning with the topic professors usually

use to introduce the subject—cost behavior. Cost behavior is anchored to variations in output; either costs vary strictly proportionately with output or they don't vary at all with output. On that dichotomy—variable versus fixed costs—hangs most of the rest of what is taught in management accounting. It underlies breakeven analysis, product costing, flexible budgeting, responsibility accounting, short-term decision problems, and more.

I believe management accounting professors anchor their subject (i.e., modern "decision-useful" management accounting) to output because they use the economists' profit-maximizing model of the firm to define that subject. In that model output is the independent variable—cost, revenue and profit are dependent variables. The economists' model is a "production function," to use the mathematical expression, that links consumption of inputs to the rate of output. The ultimate phenomenon being studied in that model is neither the managed organization nor its customers and employees. There are no people in that model, only coefficients of variability.

Cost behavior and revenue generation are, of course, more complicated in a real-life managed organization than is depicted in the economists' model of the firm. And the difference is not merely due to the model's high level of abstraction. It is due to the model's complete irrelevance to understanding behavior in a managed organization. As I said before, economists create an abstraction called the "firm" in order to construct their model of market price determination, not to understand behavior in the world's managed organizations.

When management accountants use that abstraction and bring its emphasis on output into the subject they teach, they develop information concepts that can impede competitiveness in a real business firm. Perhaps the most damaging of those information concepts are "fixed cost" and "breakeven volume." In management accounting courses, the concept of costs that do not vary with output in the short run is linked inexorably with the necessity to produce output in large volumes in the long run. Although professors never put it this way, a company could produce economically in lot sizes of one if all its costs were strictly variable in the short run. The presence of "fixed" costs presumably necessitates mass-producing quantities in large batches. The question "how large?" is answered by knowing how large are the fixed costs. Each period a company must produce at least enough units to earn margin sufficient to recoup its fixed costs (including profit, etc.). Students are taught breakeven analysis to learn how to identify the relationships between output volume, profit, and fixed costs.

Breakeven thinking promotes an insidious mindset in real-life managed organizations. It promotes the traditional emphasis on using "scale economy and high volume" to beat competitors on cost. As I mentioned earlier, however, *operating strategies driven by that mindset tend to make necessary the nonvalue activities whose costs we usually refer to as "overhead."* Most fixed costs are synonymous with "overhead" costs. Therefore, the breakeven strategy—cover fixed costs by increasing volume—resembles the dog chasing its tail. The strategy tends over time to increase the "fixed" costs it is trying to recoup—nonvalue overhead. Adhering to the breakeven strategy may be a major cause of the "overhead creep" that has plagued countless American businesses in the last two decades!

The way out is to stop viewing any costs as fixed. Instead, teach students to focus attention on activities that cause costs. Teach them how companies can empower workers to discover and capture untapped opportunities to satisfy customer wants. Above all else, teach them to think as global competitors who produce only what customers want, when they want it, with as little waste and delay as possible. In other words, teach them to think of conditions that meet the needs of people, not in terms of conditions that meet the needs of output-driven cost functions.

The emphasis educators put on "the product" has unfortunate consequences when translated into practice. Most unfortunate is the tendency it has to divert companies from giving all the attention they should to being flexible. American companies seldom manifest the global competitors' obsession with carefully studying and rapidly adapting to change in customer wants. Inertia causes them to cling tenaciously to products that too often are accidents of history, not the result of conscious efforts to satisfy the customer. Contributing to that inertia is the idea that *products* recoup costs and *products* earn profits.

Contrary to the economic model underlying management accounting education, global competitors know that products don't cause costs and products don't cause revenue. People cause costs by their activities that consume resources. And people cause revenue by creating value in the output of their activities. *Rather than focusing on the product, global competitors focus on the process of change—to hasten the speed with which they introduce new ideas to the market.* Management accounting educators must convey to students the message that to control costs and value in the global economy, managers must *lead people and manage activities;* they must *stop trying to control costs by managing output.*

2. *Using financial planning information to control operations.* The ultimate goal of business is to make money. Therefore, it is natural for top executives to view a company's operations in financial terms. That is the purpose of financial budgets with which they plan and coordinate those operations. But something is lost, I believe, when top managers roll budgeted financial goals down into the organization to establish performance targets for operating people. This has been common practice in American corporations since World War II. Teaching that practice comprises a large part of the total curriculum in management accounting.

The practice of using financial budget targets to control operations betrays the obsession with finance that many authorities now believe crippled American business in the 1960s and 1970s. That obsession was manifest in the belief that companies maximize their wealth by managing financial variables. Top managers in all too many cases saw the company as a portfolio of assets whose worth hinged on balancing risks and return, not on creating value for the customer and empowering employees to find and eliminate nonvalue activity. Quality, flexibility, and service took a back seat to issues such as debt leverage, diversification, and conglomeration. That trend led generally to the practice of managing "by the numbers," which meant top managers delegated their bottom-line financial goals down as far into the organization as possible.

Global competitors have learned in the last decade to view operating activities—not finance—as the strategic source of a company's worth. The best American companies knew this 50 to 100 years ago. The Japanese did not invent the idea—they simply rediscovered it, in many cases by reading works of earlier American authorities such as Frederick Taylor, Henry Ford, Frank Gilbreth, and, more recently, Edwards Deming, Joseph Juran, and Douglas McGregor. American management accounting educators who wish to teach a subject that is relevant to the needs of businesses must also rediscover this idea.

New information, or instinct, is needed to reinforce the tactics a company must pursue to be a global competitor. Global competitors, I believe, do not use information either from financial budgets or from financial cost records to evaluate and control a company's operating performance. Financial information, especially cost controls, seem always to drive actions in the wrong direction to achieve global competitiveness. Proper control information will drive workers to take only actions that increase value and eliminate waste. Very little has been written about this type of information. Indeed, it is not even certain, as I said before, that information as such is important to achieving global competitiveness. Instinct for "doing the right thing" may be far more important.

But companies that rely on "instinct" to drive operating activities will still need information about activity outcomes. Much is being written about this type of information—indicators of waste and value. Such information includes measures of space utilized, distances things travel, the time it takes to do things, reject rates, customer responses, total costs, and much, much more. Tracking such information on large charts kept in conspicuous places reinforces workers' efforts to do what is necessary to find waste and value. Management accounting educators should acquaint students with the logic of using these various measures and should themselves conduct research in the connections between such measures and financial outcomes such as cost and profit.

When it comes to acquainting students with the instincts one needs to do the right thing to be competitive, management accounting professors may feel they are stretching too far out of their traditional domain. The feeling is understandable. Management accounting professors are not ordinarily asked to teach techniques for identifying the wants of customers, for achieving uniform load balance, for reducing changeover times, for plant layout, for reducing time from design to market, and so forth. However, professors are being asked in this case to do the same thing as workers and executives in competitive companies all over the world are being asked to do—span boundaries. It becomes easier to address this need if management accounting professors meet regularly with colleagues in other business disciplines, especially production and operations management, and if they meet regularly to discuss problems with executives and workers in local companies.

Perhaps these efforts will teach professors themselves one of the most important instincts known to global competitors—continuous improvement. I believe continuous improvement requires, above all else, that people be given every opportunity and incentive to communicate across boundaries. In a business there are vertical boundaries (often referred to as silos) that separate functional specialties such as finance, marketing, accounting, production, engineering, and so forth. There also are horizontal boundaries that separate workers who add value from the customers they serve. Continuous improvement both requires and consists of eliminating such boundaries.

Professors must eliminate two things from the current management accounting curriculum that tend to impede continuous improvement in businesses. One is the emphasis on integrating financial planning with control information and the other is the principle of constrained optimization. Integrating plan with control information makes it *seem* possible to base financial control targets on planning goals, an insidious practice that I discussed above. Perhaps the most damaging aspect of this practice is the belief it engenders that financial results impound the most important information one needs to identify and monitor what it takes to be competitive. This belief is implicit in the use of variances from "flexible" standard cost budgets to control operating activities. Such variances impede continuous improvement by limiting managers' attention to what it takes to achieve targets that are static and merely financial. Global competitors never stop reaching and they never limit their reach to financial variables.

My criticisms of using planning information to control performance does not mean, however, that management accounting professors should stop teaching students how to budget. On the contrary. The goal of business is to make money. To achieve this goal, top managers must be able to project the likely financial consequences of actions they plan to take. Equally important, they must continually track the cash consequences of actions as they occur. Competitive organizations can go bankrupt if someone is not watching the lockbox. But the information in budgets that is necessary to achieve financial soundness and solvency is not sufficient to insure competitive operations.

Constrained optimization is the other item in the curriculum that is antithetical to the philosophy of continuous improvement. Constrained optimization is implicit in the economists' theory of the firm. It is impossible to find "optimal" outcomes at stable equilibrium points without assuming at least some constraints, trade-offs, or limits. By definition, "optimal" outcomes in the economists' model are static and unchanging until something happens to change the constraints. A good example of this idea at work in the management accounting curriculum is the "economic order quantity" (EOQ) paradigm. The EOQ paradigm defines production lot sizes that minimize total costs of two constraints—setup time and inventory holding costs. Adhering to EOQ lot-size calculations does not encourage managers to produce what is needed, when it is needed. Indeed, companies that pursue traditional "scale economy and high volume" practices can show with EOQ calculations that it is uneconomic to produce just up to the rate of customer demand and then stop. Meanwhile, their global competitors can show how reducing one of the constraints— setup time—allows one to produce lot sizes as small as you want. As a general rule, the path to global competitiveness is achieved by moving constraints, not by optimizing costs within constraints.

CONCLUSION

Professors will not shift the focus of university management accounting programs from their traditional academic customer to managed business organizations unless they first understand how changes in the global economy since 1970 render traditional management accounting virtually useless. Those changes affect management accounting through the radically different practices businesses must adopt to be competitive in the global economy. Until 1970 American businesses competed on cost by seeking scale economies and high volume business. The desired state was to be mass-producing long runs of homogeneous output with plenty of inventory to hedge against poor quality and inflexibility. Now companies must compete on giving value to the customer by achieving continuous and rapid flow of value-adding activity. Today's desired state is to be producing a large variety of things in small lots with inventories kept as close to zero as possible. Management accounting in the old order emphasized costs and variances; in the new order it will stress value to the customer and waste in activities.

In the global economy, companies that persist in controlling unit costs by managing output will watch their profits fall as value diminishes and total costs rise. Only companies that manage value and waste will see profits rise.

Global competition has several profound implications for the content of management accounting courses taught in universities. First, it shifts the central focus of attention from product output to value in activities. This shatters the dichotomy between fixed and variable costs that is central to traditional management accounting teaching. In turn, breakeven analysis ceases to be a useful concept. Secondly, global competition leaves no place for delegating financial planning goals to operating personnel. This eliminates from the curriculum most of the material on financial control systems, especially the use of variances from standard cost budgets to monitor performance of operating personnel. Finally, global competition calls into question the concept of constrained optimization.

These changes in management accounting education might seem a tall order. Indeed, they are. But time to implement such changes is growing short. As businesses alter their practices to compete in the global environment, they will become acutely aware of the gap between their management techniques and obsolete knowledge being taught new university graduates. Rather than spending resources to retrain new dogs who have been taught old tricks, businesses may begin increasingly to hire graduates of nonbusiness university programs and train them from scratch themselves. One sees this already in the increased attention businesses give to graduates of good liberal arts institutions. From a broad perspective that is probably a welcome trend. In universities with business programs that cling tenaciously to the traditional curriculum, potential management accounting students may see this trend and adapt by registering in nonbusiness programs. The only people who are certain to lose out if this happens will be professors of management accounting in programs that fail to adapt to the global environment.

SUGGESTED ADDITIONAL READING

On the role of information in activity management:

H. Thomas Johnson. "Let's Return the Controller to Relevance: A Historical Perspective." *Cost Accounting for the '90s: Responding to Technological Change* (National Association of Accountants, 1988). pp. 195-202.

————."Activity-Based Information: A Blueprint for World-Class Management Accounting." *Management Accounting* (June 1988). pp. 23-30.

————."Activity-Based Information: Accounting for Competitive Excellence." *Target* (Spring 1989). pp. 4-9.

————."Managing Costs: An Outmoded Philosophy." *Manufacturing Engineering* (May 1989). pp. 42-46.

————."Performance Measurement for Competitive Excellence." *Measures for Manufacturing Excellence*, ed. by Robert S. Kaplan (Harvard Business School Press, 1990). In press.

On management accounting education:

Thomas R. Dyckman. "The King's New Clothes?" *Accounting Horizons* (June 1988). pp. 115-122.

————."Practice to Research—'What Have You Done for Me Lately?'" *Accounting Horizons* (March 1989). pp. 111-118.

H. Thomas Johnson and Robert S. Kaplan. *Relevance Lost: The Rise and Fall of Management Accounting* (Harvard Business School Press, 1987). chs. 1, 6, and 7.

J. S. Jordan. "The Economics of Accounting Information Systems." *American Economic Review* (May 1989). pp. 140-145.

IMPROVING PERFORMANCE THROUGH ACTIVITY ANALYSIS

Thomas O'Brien
Manager, Production Management Programs
General Electric Co.
Bridgeport, CT 06602

ABSTRACT

This paper discusses the concept of Activity Analysis as a process to increase productivity and performance within an organization and its sub units.

The goal of Activity Analysis is to identify activities performed by the company, analyze them in terms of their value to the organization and to perform the activities deemed necessary in an efficient manner.

The framework with which to use this technique includes defining value in terms of stakeholder satisfaction and translating that value into concrete measurement factors. Characteristics of activities of value are timeliness, efficiency, comparability to the industry leader, and correctness in application.

The analytical process should clarify such issues as cause and effect, priority status, duplication, and comparability of the activities studied.

The end result of Analytical Analysis is the implementation of more efficient operating procedures that meet organizational goals at minimal expense.

From 1981 to 1987 the GE Company has grown its earnings from $1.7 billion to $2.9 billion, an average annual growth rate (AAGR) of slightly better than 9%. During this period, the GNP grew at an AAGR of 2.9%.

One gross measure of productivity, Sales per Employee, signals how part of this was accomplished. During this same period, sales increased from $27.2 billion to $40.5 billion and employment fell by approximately 100,000. This resulted in Sales per Employee doubling from $67,000 to $134,000, an AAGR of 12.2%. When inflation was removed, the AAGR fell to 8.7%. This was accomplished both by increasing the numerator of the productivity equation (output divided by input) through acquisitions, joint ventures, internal business development and growth, etc.; and, by reducing the denominator through divestitures, better understanding and control of base costs, sourcing more components and products, etc.

Our Chairman has stated to the investor community that the Company plans to grow its earnings at from one-and-one half to two times the GNP. One major effort that will support this growth prediction is our continuing challenge to increase productivity.

GE plans to obtain this productivity improvement through efforts on many different fronts, e.g., better product designs, workforce involvement, process improvement, etc.

There are many additional efforts within the Company to improve productivity. This paper will focus on how Activity Analysis can be used to improve productivity and will demonstrate how two locations used it to improve productivity in their business. Since very little in GE is legislated from the Corporate Office, there is no "Thou shall do activity analysis" command from on high. The Corporate Staff has the responsibility for identifying best practices, such as activity analysis, and transferring that knowledge to the operating businesses. It's up to them whether or not they wish to do something about it.

WHY ACTIVITY ANALYSIS?

Performing the right activities, and performing those activities in the right way, can affect the way a business satisfies its stakeholders. And, by stakeholders is meant not only its customers, but its stockholders and its employees as well. Satisfying stakeholders can mean larger market share, greater profits, higher productivity, and so on. And, activities, depending upon how fine they are granulated, can be the basic element of people effort within a company.

The objective of this paper is to demonstrate that a business can *improve* its performance and move towards excellence by *identifying* what activities are operating within its walls and by systematically *analyzing* them so that action can be taken to eliminate the unnecessary activities, to strengthen the necessary activities, and to perform the necessary activities in the most efficient way possible. This paper will further demonstrate how first to identify what activities are being performed; second, what actions to take, that is, how to act on this new knowledge of activities; and finally, will give some examples of improvements that have been made as a result of activity analysis.

This paper discusses activities and activity costs. It does not present a systematic way of continually capturing these costs into an activity based costing system. The experiences depicted will identify activities, their associated costs, and the actions that needed to be taken, at a specific, single point in time.

Much has been written lately on value added and non-value added activities. In our analysis of activities, needless to say, both have been found. Of course, activities that do not add value to the business beg the question, "Why do them at all; how can we get rid of them?" Activities such as approving purchase order requests could fall into this category. And, even though an activity might be considered value added, it has been found that in some businesses certain activities can be unknowingly duplicated, that is, done unnecessarily in more than one place in the organization. For example, while product pricing might typically be the responsibility of the Marketing organization, if one were to find that Customer Service was also providing prices, it would indicate that there was something wrong with the system, and that this duplication could lead to confusion.

Activities are often fragmented, i.e., done in a number of places when, in fact, they should be focused in one area so that they can be more effective. We have found businesses that knowingly permit purchasing of supplies to be done not only by the Purchasing personnel but also by engineers, maintenance people or others. While in some cases there may be a legitimate reason for doing so, it is usually more efficient to focus that purchasing effort in one organization.

Also, while activities may be value added, they may be unnecessarily sequential. We have found cases where value added activities are done in series, one after another, and by so sequencing them, took an inordinate amount of time to perform the entire series; in reality, a number of these activities could be done in parallel, significantly reducing the amount of time it took to perform the set of activities. A perfect example of this is the work that is being done in many factories in reducing the amount of time it takes to set up processes by performing many activities offline.

Next, activities can be costly. While the activity itself might be necessary, it is not being done in the most efficient manner. An example of this might be where a buyer places individual orders with a supplier, when a blanket order covering the whole year would be a much more efficient way to accomplish it.

Finally, activities may be performed contrary to the objectives of the business. A business, for instance, could set priorities for customer service, e.g., customer A will be treated better than all other customers when the activity "order entry" is performed. If this is not sufficiently communicated to the clerks that interface with customers, others may be getting the special treatment that A should be receiving, at A's expense.

AN ACTIVITY ANALYSIS FRAMEWORK

All this suggests that there should be some rules for analyzing activities. The following framework for this activity analysis is suggested:

- For an activity to be of value it must satisfy some stakeholder, be that stakeholder a customer, a stockholder, an employee, or other.
- Satisfying stakeholders must be translated into specific critical success factors for that business, e.g., if ROI is critical to how a company is valued in the market, then the broad area of inventory planning might be key and a specific activity such as "stockroom control" critical. Therefore, activities need to be screened for relevance to a specific business need or critical success factor.
- An activity must be able to withstand comparison to the best practice for that activity. If a necessary activity is being performed in an inefficient manner, that activity should be highlighted for potential improvement. For instance, the activity "taking customer orders" is certainly necessary and may be considered value added. But, if it is done manually when the best practice suggests that it can be done through EDI (electronic data interchange), then one might question whether or not the way it is being done today is truly value added, and how much potential for improvement there exists in that method.
- Search for the best linkages of activities to reduce time and or satisfy our stakeholders. Often organizational barriers inhibit the linking of activities to improve business performance. An oft used example is how the Japanese have linked their activities in Marketing, Engineering and Manufacturing together to improve the time for new product introduction.
- Of course activities must be performed in a timely manner. The value added activity that is done late may be worthless to the business.
- One must have a way of prioritizing and concentrating on the important few activities that mean the most to the business. We have found that there are usually between 200 and 300 activities that reasonably describe what a business is doing. It is nearly impossible to concentrate on all of those activities.
- Finally, performing an activity correctly should optimize total performance and encourage teamwork amongst the members of the business. This is likely to happen as a by-product because the business has tied all of its activities to a shared set of objectives.

Two cases will be described, both of them from the GE Company. One is a small shop; the other a total business. In each case the approach used to identify the activities that were taking place, the method or methods used for analyzing those activities, and the types of improvements that were gained by each, will be discussed.

SHOP CASE

This function is basically an electromechanical assembly shop with low volume (approximately ten units per week), and the product is of high value (thousands of dollars each). It is a high tech product. The problem in this shop was that though it had tried to implement a number of just-in-time concepts and had succeeded up to a point, their performance was still sluggish relative to what they could attain if they fully implemented more of these concepts. They approached the problem by identifying the activities that were being performed in their unit, analyzing those activities, and attacking them so that overall performance could be improved.

The results of their activity analysis program were dramatic. Productivity was improved by over 20%. Cycle time and inventory was better than cut in half and quality improved as the number of defects plummeted. Finally, their customers were more satisfied not only because of improved quality, but also because they were able to improve on their schedule attainment. The following paragraphs describe how they went about this activity analysis.

The first step was to identify all the activities that were associated with planning, operating, maintaining, and improving the process in their, what they began to call, Value Added Activity Center, or VAAC for short. Some 30 activities were identified. Fig. 1 shows eight of those 30 and the number of equivalent people that were performing each activity.

The manager of this VAAC and his staff essentially brainstormed both the activities performed and how each person associated with the unit was spending his or her time. As you can see, when they added up all the activities it amounted to 31.5 equivalent people. The next step was to determine the value of these activities.

Figure 1. Workforce Activity Distribution

Activity	Equiv. People	Value Added	Gray	Waste
Assemble	10.7	10.7		
Accumulate Material	3.9			3.9
Expedite Material	2.1			2.1
•				
•				
•				
Work Assignment	1.2		1.2	
Wait for Material	1.2			1.2
•				
•				
•				
Implement ECN's	.3		.3	
Empl Communication	.3		.3	
MRB	.3			.3
Total	31.5	10.7	4.7	16.1
		34%	15%	51%

They further categorized these activities into value added, gray, and waste. The gray category meant that they were undecided as to whether it was a value added or non-value added activity. The only certain value added activity was the first activity of "assemble" and this amounted to only 34% of the total activities performed in the VAAC. On the other hand, the waste activities amounted to better than 50% .... a great opportunity if part or all of that could be reduced or eliminated.

Across the top of Fig. 2. are the five highest waste activities: accumulate material on the floor, expedite material, move material, rework, and test and verify. In a storyboard, brainstorming session, the same team mentioned before identified what the drivers of each of these waste activities were. The drivers of accumulating material on the floor were things such as the way the stock was laid out, the procedures in the stock room, and the assembly sequence on the floor. Also, the volume of part numbers drove this activity—the more parts, the more difficult and time consuming was the accumulation of material.

The third activity, move material, reveals that the way the shop is laid out, the equipment used for material handling, and the number of labor classifications were all drivers of this waste activity. Notice that the team segregated the drivers into three categories. Internal drivers were those the VAAC could handle themselves— they could make the changes and effect improvements in their performance. There are also external drivers. These were the drivers over which they did not have direct control but did have some influence. And finally, there were some semi-givens, those things which affected the waste activity but were somewhat fixed, at least in the short term.

Figure 2. Five Highest Waste Activities

Activity	Accum Mtl (Floor)	Expedite Material	Move Material	Rework	Test/ Verify	Total
Equiv. People	3.9	2.1	2.1	1.9	1.9	11.8
Internal Drivers	• Stock Layout • Stocking Procedures • Assembly Sequence	• Scrap • Stock Errors • Ordering Errors • MRP Lead Times	• Layout • Handling Equipment	• Asm Errors • Large Batches • Damage • Methods/ Procedures	• Interpretation of Design • Training • Quality Problems	
External Drivers		• ECNs • Schedule Changes • Supplier Delivery & Quality		• ECNs • Supplier Quality		
Semi-Given	• Volume of Part Nos.		• Labor Classification	• Quality of Design	• Design • Regulations	

As the team scanned this matrix, they determined that the one major driver across the whole VAAC was the layout and flow of material. They attacked these drivers and, in so doing, were able to make the significant changes in performance mentioned previously.

BUSINESS CASE

In the previous shop case, the number of people was about 30. In this next case, an entire business of several thousand people is used. This business is a heavy mechanical/electrical assembly type. Volume is low ... several hundred items a year. It had high base costs, and this became a very critical item when volume began to drop off. Lead times were long, sometimes 18 months for special items. This business was number two in its market. It was a cyclical business and was on the declining portion of that cyclical curve. In an attempt to reduce its base costs, the business embarked on an activity identification, analysis, and improvement program.

As a result of this program they were able to remove one layer of management; took out about $20 million (the majority of that coming from base costs); and because of this cost reduction, plus product and process improvements, they became number one in the marketplace. The following describes the steps they took in their activity identification, analysis, and improvement.

After they had defined their objective of lowering base costs, their first step was to develop a dictionary of activities that contained approximately 250 activities which fully described all the activities within the business. A team of about six people from that business started with a generic dictionary of activities and modified it to suit their business.

Depending upon the objectives of a project, and specifically what activity information is needed, the activities could be defined more grossly or in more detail than was done in this case. For instance, the activities of Application Engineering-Distribution and Application Engineering-Original Equipment used in this case, could have been described under a single activity called "Application Engineering" if another business was not interested in differentiating the effort being applied in the original equipment market, vs. the distribution market. On the other hand, Application Engineering-Distribution could have been even further detailed to identify the activities by product line, if that was important to a business in its future analysis. The activities in the dictionary are listed alphabetically and not by functional area to insure that the person scanning the dictionary, an engineer for instance, would not just go to the Engineering Section and only consider those activities. We often find that people perform activities outside their own functional area.

The next step is to identify who is performing what activities. To do this, we ask each manager to collect that data for all the people that are reporting to him.

On a preprinted standard form, each manager lists the names of the people reporting to him, their job titles, identification number, annual earnings, and the percent of time each applies to the activities that are performed in his organization.

With the data now captured, the next step is to enter this data into a relational data base so that it can be manipulated for whatever analyses need to be performed. The following sections describe types of analyses that are typically performed on this data in order to identify opportunities for improvement.

ANALYSIS #1 - 80/20

The first analysis is called the 80/20, or pareto, analysis (see Fig. 3). It merely ranks from top to bottom the most costly to the least costly activities. The activity "manage" amounts to the highest cost in this business, a total of 16.9% of total payroll. It is also noted that 45 people have indicated that they are performing this activity but obviously not all full time, since there are only 18.9 equivalent total people spending full time managing. This simple analysis permits the business to quickly see which activities might need to be attacked first.

	Figure 3. Analysis #1 - 80/20				
Activity	Actual People	Equiv. People	Dollars	% Total	Cum %
Manage	45	18.9	463,480	16.9	16.9
Selling	33	11.9	130,725	5.1	22.0
Secretarial	21	12.2	123,940	4.5	26.5
Expediting	25	5.7	109,270	4.0	30.5
.	.	.	.	.	.
.	.	.	.	.	.
.	.	.	.	.	.
.	.	.	.	.	.
.	.	.	.	.	.
.	.	.	.	.	.
.	.	.	.	.	.
.	.	.	.	.	.

ANALYSIS #2 - CAUSE AND EFFECT

This second analysis called cause and effect, uses the previous 80/20 analysis, which identifies those few activities which account for 80% of the cost, and compares them to the many activities that account for a small amount of cost to determine if there are any cause and effect relationships. In this case, it was noticed that the activity "grievances" costs $160,200. On the other hand, only $6200 was being spent on "employee communication." Would more effort put into employee communication have a leveraged payoff by reducing the amount of dollars associated with grievances? This type of analysis could highlight that opportunity.

ANALYSIS #3 - PRIORITY/COST FIT

Since the top 20% of the activities that account for 80% of the cost have already been selected, the general manager then prioritized this set of activities from the highest to the lowest and placed them in four groupings based on where he thinks the most effort, energy and cost should be spent. The next step was to compare the general manager's priorities with what the actual priorities (expenditures) are.

Those activities that are along the top left to bottom right diagonal are in sync. In other words, the general manager believes that the business should be spending high on the "selling" activity and the actual cost associated with selling corroborates that. This analysis also suggests that those activities that are in the upper right hand corner such as "pricing product, telephone inquiries and operation services," where the costs are higher than the priorities set by the general manager, should be reviewed for possibly being too high. On the other hand, those activities in the lower left hand corner, "inventory control, finished goods, field failure/marketing and production planning" might be considered too low, possibly needing more expenditures in order to better accomplish the objectives of the business.

Figure 4. Analysis #3 - Priority/Cost Fit

General Manager's Priorities

Highest - 1	2	3	Lowest - 4		
Selling	Train Cust. Buying Influence Contacts	Pricing Project	Telephone Inquiries	Highest 1	A c t u a l
Distribution Development	Selling Svs. Sales Plan Implement.	Application Svs.	Operations Svs.	2	E x p e n d
Inv. Control/ Finished Goods	Merchandising Svs.	Train Salesmen	Mkt. Analysis Salesman Report	3	i t u r e s
Field Failure/ Marketing Prod. Plan		Merchandising Planning Sales Plan Form	Pricing Sales Order Processing	Lowest 4	

(Right columns labeled: Actual Expenditures, with scale Highest 1, 2, 3, Lowest 4)

ANALYSIS #4 - MISPLACED OR FRAGMENTED

Fig. 5 shows an example of a business with five organizations. For the first activity "pricing," by far the greatest amount of effort occurs in organization number one, but a small amount, 15%, is done in organization number four. Organization number one is Marketing, which is supposed to have sole responsibility for pricing,

Figure 5. Analysis #4 - Misplaced or Fragmented

% of Total Activity Performed in Organization

Organization	1	2	3	4	5
Activity					
Pricing	85%			15%	
Customer Training	60%		20%	15%	5%

and organization number four is Customer Service. Are the people that are doing pricing in Customer Service doing it by default? Are they issuing the right prices? Are they in conflict or in concert with the price policies of Marketing? This type of analysis surfaces these types of questions to be answered.

The second activity "customer training," reveals that four of the five organizations within the business perform this activity. The mission of each organization had training specifically written into it, some more than others. The questions that might arise here are "Is the customer training too fragmented?", "Could a better job be done if the training expertise were consolidated into one, or maybe two, organizations—thus providing the customer with higher quality training?"

ANALYSIS #5 - PRIMARY VS. SECONDARY

On the input form mentioned previously, under each of the activities identified as being performed in an organization, the manager would indicate whether that activity was primary to the mission of his organization, or secondary. With that in the data base, a summary can easily be made for each organization, re how much of their total effort is primary to their mission and how much secondary. An organization with 95% of the activities primary would be well focused on its mission, and one with only 50% of its effort directed at its primary mission would not. Based on experience, about 80% primary is the minimum amount that a well focused organization should be spending on its primary activities.

The data base permits us to compare like organizations within a business. In Fig. 6 there are four comparable Engineering organizations: Mechanical, Control,

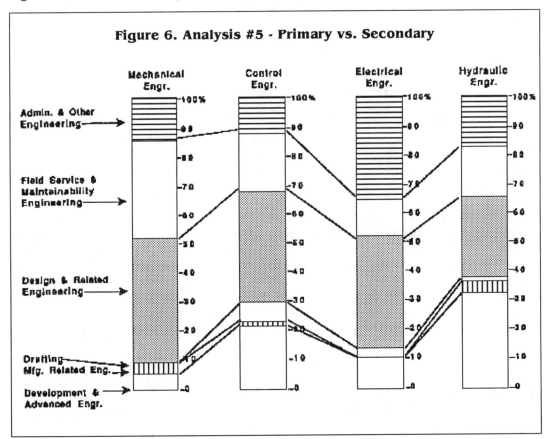

Figure 6. Analysis #5 - Primary vs. Secondary

Electrical and Hydraulic. It was known to virtually everybody in the business that Control Engineering far outshone the Electrical Engineering group ... they met their commitments, on time, and the quality of their work was superior. In both these cases, the first activity "administration and other" was considered a secondary activity; the other activities beneath that are all primary. It is little wonder that Control Engineering was doing the better work, 90% of its efforts was directed at its primary mission. On the other hand, in the Electrical Engineering group, about 65% of its efforts was directed at its primary mission. This type of analysis could suggest deemphasizing certain activities or changing the mix of personnel in an organization to concentrate more on the primary activities and thus better meet the objectives of that organization and the business itself.

ANALYSIS #6 - COMPARISONS

This last analysis "comparisons," compares the amount of effort being applied to an activity to that applied in similar businesses. If, for instance, a business spends 1.1% of its total payroll on training, and other similar businesses in our data base spend significantly more, the questions to be raised might be "Are they spending enough on training?", "Would additional funds spent on training help us reduce other costs and/or better meet the objectives of the business?".

Conversely, in another situation, a plant might be spending 4.7% of its payroll on material handling, and this was inordinately high when compared to other similar businesses. The question to ask here could be, "Would a better layout or automated equipment reduce material handling costs?" Again, just as with all of the other analyses, there are no immediate answers but questions are raised which demand further investigation.

DIGGING FOR DRIVERS

In trying to understand the drivers of activities, three examples are given of how digging for drivers led to important changes and improvements in this business.

The "timekeeping" activity represented a significant expenditure, $1.2 million. It was found that approximately 80,000 vouchers were processed each week, and this was driven in part by the 11 wage payment instructions and by the nine expense account categories. To reduce this amount of timekeeping effort, the business modified their accounting system to eliminate all but two expense accounts and simplified the administrative process. They also established a single day work rate for each incentive worker and used one voucher a week for all expense payments. These actions took out more than one-third of the timekeeping costs.

In a second example it was found that while $2.5 million was being spent on "cost reduction" activities, the effort in this business was only 60% of that in comparable businesses. Further analysis showed that the effort being applied to cost reduction by Producibility Engineering was critically low when compared to other businesses. Experience suggests that probably 70 to 80% of the product cost is determined during the design stage when Producibility Engineering plays a major role. Beefing up that organization was expected to yield cost reduction results in later years, and it did.

Our final example combined the set of activities identified as "analysis, planning and reporting." Here the business was spending $1.7 million annually. This was high compared to other similar businesses, and further investigation revealed that there were 100 measurements in Manufacturing which were reported and analyzed periodically.

Upon investigation of both the need for certain of these measurements and the actions which they produced, it was found that several of the measurements were counter-productive, and drove hidden costs. For instance, one of the primary measurements on a foreman was direct labor utilization, and the higher the better. This drove him to use personnel to make parts and assemblies, whether or not they were needed, thus unnecessarily increasing inventory. The solution here was to eliminate or change a number of the measurements, reducing costs to about two thirds of what they had formerly been.

SUMMARY

In both the shop and the business case, the steps were essentially the same—identifying the activities, analyzing them, and going a little deeper to understand their underlying drivers; and, finally taking action to improve the business. In the case of the shop, the identification of the activities was rather informal. They used a brainstorming or storyboarding approach to identify their 30 activities. It was all done rather quickly, in a day or so.

On the other hand, in the larger business case, the development of a very precise dictionary, coupled with a very structured approach to gathering data on the 250 activities, plus the computer processing, took almost two months. In the analysis phase, in the shop, the approach was hands-on problem solving. It was easy enough for a small group of involved people to get their arms around the 30 activities associated with that shop. In the business case, there were a number of prescribed analyses that were performed. Our target business was compared to other businesses to identify discontinuities. An 80/20, pareto, analysis of the activities was performed to determine which ones offered the most opportunity. The general manager's priority of activities was compared to the actual cost of activities to determine where there might be opportunities to reduce or increase resources. Primary work was segregated from secondary work, and activities were reviewed in terms of whether they might be misplaced or fragmented.

The improvements made in each case differed in time and in scope. In the shop, the actions were taken within weeks and the improvements followed quickly thereafter. They realized improvements in productivity and reduced cycle time. On the other hand, in the business, the actions began six months after they launched into the program and continued for 18 months, but the rewards were significant. They increased their market share, became number one in their market, and their ROI also significantly increased.

In summary, the author would like to leave the reader with the following thoughts on activity analysis:
- Activity analysis can equal satisfied stakeholders both by helping a business to become more effective, performing only the right activities; and to become more efficient, performing those activities in the right way.
- A framework was proposed which can serve as a guide as a business launches into its program of activity analysis, and finally,
- Whether a small shop or a large business, remember the three steps necessary on the road to satisfied stakeholders: *identify* your activities, *analyze* them using the framework as a guide, and take action on the drivers to *improve* your operation's performance.

QUALITY FUNCTION DEPLOYMENT IN MANUFACTURING AND SERVICE ORGANIZATIONS

Rene Quevedo
Ernst & Whinney
1400 Pillsbury Center
Minneapolis, MN 55402

ABSTRACT

Quality Function Deployment is a planning and management tool whose goal is to identify customer value in terms of process value so as to meet customer demands as efficiently as possible.

Several analytical techniques that identify customer wants in production terms, the level at which customer wants are satisfied and conflicts between production goals are used to assess the adequacy of an organization's quality function. The "House of Quality" concept displays the interaction of these analyses in concise graphic format.

Quality Function Deployment occurs at progressive levels with each step defining a goal and a method of achieving it. Macro level planning of customer wants is first identified, followed by product design, product specification and process control.

QFD is effective only if adequate market research is conducted and must be adjusted to changing market needs.

Few concepts have caught the attention of corporate executives as much as Quality Function Deployment or QFD has. Originally an engineering "tool," it is fast becoming a force driving process designs, from manufacturing to utilities, from insurance companies to banking institutions. Its applicability is becoming widespread, from product design to software engineering, streamlining of administrative processes, systems downsizing and overhead cost reduction, just to name a few.

First articulated and utilized by Mitsubishi's Kobe Shipyard in 1972, it was not until 1977 when the results obtained by Toyota began placing QFD as a powerful planning tool. In 1978, professors Akao and Mizuno published a QFD text in Japan, and in 1979 Toyota incorporated its suppliers into the program. Komatsu (WA project) and Dymic Corporation (design of cars on the Japanese National Railway's Shinkanseu "bullet" train), reached notoriety in 1982.

In the U.S., QFD activity began in June 1984, when Dr. Don Clausing (formerly with Xerox, now with the Massachusetts Institute of Technology) introduced the concept at Ford Motor Company. The American Supplier Institute (ASI) organized study missions to Japan in 1984, 1985 and 1986 to evaluate Toyota's experience. Pilot projects at Omark Industries (1985), Budd Engineering Company (1986) and Kelsey-Hayes Co. (1986), also represented a significant contribution to the body of knowledge. Another key influence has been Bob King (GOAL, Lawrence, Massachusetts) whose close relationship to Dr. W. E. Deming and Dr. Joji Akao (Tamagawa University, Japan, considered the "father of QFD") has greatly contributed to the acceptance of the tool as part of the total Quality Management process.

WHAT IS QFD?

Simply stated, Quality Function Deployment is a process through which customer requirements are translated into design features, process specifications and operational targets. This way, customer-perceived value *drives* outcome and output requirements, and every activity throughout the value chain. QFD is therefore a planning, communication and process documentation system.

STRUCTURE OF QFD

The House of Quality is the "heart" of QFD analysis. The term was popularized by Larry Sullivan and William Eureka (American Supplier Institute, Dearborn, Michigan). Fortuna (1988) describes this framework with an example: the starting point for QFD is a positive statement of what the customer wants and needs. In other words, what are the project objectives or what are the ends to which the company is working. These are not necessarily product specifications, but may be more general in nature.

For example, a commercial printer may tell its paper supplier that it wants no tears while the paper is running on a rotary press. However, such general requirements cannot be acted upon directly. Therefore, using the company's internal technical language, it must be determined what means will be used to accomplish the ends. At the product planning stage, these are often called substitute characteristics or design requirements (more generally in QFD, the ends and means are often referred to as whats and hows, respectively). So for this example, the customer's requirement of no paper tears can be translated into design requirements for thickness, width and tensile strength. Target values that are as specific as possible also must be determined. Width and thickness then might be assigned targets measured in millimeters, and tensile strength a target in terms of pounds of force.

Except for the simplest of products, the relationship between the ends and means can quickly become very confusing. This problem can be solved by forming a matrix from the lists of ends and means, and by showing the relationships among them by the use of various symbols. Figure 1 shows a simplified conceptual picture of this most basic QFD tool.

The matrix is, of course, a simple concept. But it is a disciplined way to compare two sets of items. It provides a logical, in-depth look at many of the critical aspects of any product or service. The focus can be placed on the customer so a widespread number of correlations and relationships can be considered. In short, it helps to ensure that things "don't fall through the cracks."

While the matrix in Figure 1 represents the basic logic of all QFD charts, there are many options and enhancements which are frequently used and can help systematically and successively apply the Pareto principle. That is, elaborate the details at one stage, then select the most important items for the next stage. Thus, QFD cannot only help determine where to concentrate engineering effort, but, just as important, where not to invest time and money.

Another option for product planning is to make competitive evaluations of customer demands and substitute quality characteristics. Ask customers how well your company fares against the competition on their most important requirements. Then ask your engineers how your company rates against the competition on the technical requirements that have been specified for the product. Taken together, these evaluations can often help pinpoint how a company might gain a competitive advantage and where improvement might be needed most.

A correlation matrix, where the substitute characteristics are compared against each other, can also be used to identify conflicting design requirements. For example, the design requirements for a diesel engine might include targets for acceleration and particulate emissions. These two requirements might have a strong negative correlation in the sense that as the emissions improve, the acceleration worsens. This type of exercise helps detect possible trade-offs early on, a primary goal of QFD.

Figure 2 is a conceptual product planning matrix that includes all of the enhancements previously described. This is often called the house of quality, and it is almost always the first chart to be completed in a QFD study. This chart identifies a handful of key customer requirements and substitute characteristics that will become the focus of the rest of the QFD study. In general, these characteristics are transferred to subsequent charts to be explored in greater detail. Figure 3 illustrates a partial matrix relating customer and design requirements for the commercial printer example.

PART DEPLOYMENT, PROCESS AND PRODUCTION PLANNING

Once the initial matrix (Customer vs. Design Requirements) is made, each Design Requirement is broken down in a second level matrix, linking each of these to parts or components of the Design (Figure 4). Each requirement becomes a "what" with relationships to processes and stages throughout the entire operation. In summary, there are at least four levels of matrices commonly used, an illustrative sequence is summarized below:

EXAMPLES (*)

MATRIX	WHAT	HOW	COMMENTS
Product/Service Planning	Easy to transport	Low total weight	• Do not over-define product • Go by what customer data indicates
Product/Service Design	Low total weight	Weight of materials	• What makes the total weight of the unit?
Process Planning	Weight of materials	Wood	• Specs (Purchasing)
		Steel frame	• Specs (Purchasing)
Process Control	Wood	• Punch 2 holes (1/2 inch) • Saw to 6' by 6" shape	• Shopfloor Specs
	Steel	• Drill 4 holes (1/16 inch) • Weld to shape	• Shopfloor Specs

(*) Partial. Used only to illustrate linkage between Tasks and Customer Value.

In addition, QFD is often associated with concepts such as Taguchi's Loss Function (i.e., for definition of target values), Design of Experiments (DOE), Failure Modes and Effects Analysis (FMEA) and Fault Tree Analysis (FTA). Once a process is fine-tuned to deliver *exactly* what customers need, Value Analysis/Value Engineering (VA/VE) tools are then used to improve such processes in order to maintain or enhance the level of performance, at a lower cost or shorter time. So in a way, QFD is only the beginning.

POINTS OF CONTACT BETWEEN CUSTOMERS AND SUPPLIERS

Perhaps the greatest contribution of QFD is the precise definition of customer value-criteria (Figure 5) as the driver of process designs. Throughout its sequence, the achievement of specific components of customer-perceived quality is linked to policies, processes, activities and tasks. By the same token, it forces scrutiny of such elements in view of their contribution (or lack of) to value.

Each point of contact between customers and suppliers (Figure 6) becomes an area for possible competitive differentiation and revenue generation. Knowing exactly what customers expect (or are willing to pay for) at each point of contact from the time they first become aware of a company or product, throughout the purchase transaction and after the sale, is paramount. Such an interaction places demands on several areas of an organization, as illustrated in Figure 7.

QFD is beginning to be used to analyze the "Total Purchasing Experience," beyond the product and service delivery arena. Again, every contact between buyers and sellers represents an opportunity for differentiation and revenue generation. The price a customer is willing to pay for a product or service is closely linked to his or her perception of value. On the other hand, knowledge of the contribution of organizational activities to the achievement of such value is vital to focus improvement.

QFD AND PROCESS VALUE

As stated before, customers associate price with his/her perception of value throughout the total transaction. But the concept of Process Value, even though driven by these same perceptions, is basically the accumulated cost of basic or value-adding tasks and activities, those contributing to what customers perceive as getting (Figure 8). The difference between Process Value and customer-perceived value is, in theory, the sum of profits and process waste. Another comparison is that of Process Value vs. Process Costs. With QFD defining a logical value chain on a product-by-product or service-by-service basis, potential enhancements of processes to maximize value are going to be evaluated on the basis of either potential revenue generation and/or cost reduction (via waste elimination).

Traditionally, estimating revenue increases to be derived from product improvements is a very difficult task indeed. Furthermore, revenues attributed to process enhancements are almost unheard of. But determining potential cost savings are somewhat easier, as illustrated in the following example.

ESTIMATING PROCESS VALUE: A CASE STUDY

Figure 9 shows a white-collar department supporting a large retail operation with 55 locations. It is composed of invoice processing, accounts payable and inventory management. Even though it does not appear to have a direct contact with the final customer, it provides critical information to store managers on stock availability and upcoming shipments, thus playing a key role on satisfying customer needs. In fact, stockouts were believed to be partially responsible for a declining market share in some markets (initial competitive assessments ranked it highest among the list of complaints). A very strong relationship between stock levels and the inventory management area puts the focus on the department as a whole. Main-frame-based Accounts Payable and General Ledger systems are part of the software in use.

Activities were then broken into tasks, identifying the percentage of time needed to complete (related to time to carry out the entire activity). As an example, Figure 10 illustrates such a breakdown for the activity "Record Journal Entries." Here, "value" is delivered when the entry is made (10% of total effort), but 60% of the time is spent finding and explaining differences. Upon Functional Analysis of the activity, it was determined that making the entry is the only basic (value-adding) function, with all others being secondary (non-value adding). Assigning worth exclusively to the value adding task ($113,066/yr.), a cost/worth ratio of 10 was determined. Thus when carrying out the activity "Record Journal Entry," this company needed to spend $10 of cost to generate $1 of value.

By summarizing all activities, basic functions, percent, costs, and worth, cost/worth ratios by activity were consolidated into a departmental index of $3.44 (Figure 11). Thus $3.44 of cost were incurred to produce $1 of value. In other words, $2.44 of waste were generated per every dollar of value. Thus of $1.6 million/year of departmental cost, only 29% was value. Figure 12 plots value vs. cost along the value-adding process, clearly identifying accumulated waste. Most importantly, however, addressing waste-generators and other cost drivers was considered critical to achieve an important customer requirement, no stockouts. In the case of the Journal Entries, differences between invoiced and received amounts were traced via statistical analysis to Price Differences (pricing file of A/P system), and Discount Policies (bulk purchases/procedure-related).

By analyzing value and non-value adding tasks, road map for improvement was formulated (Figure 13). By focusing on eliminating/minimizing secondary or non-value adding tasks, and speeding basic or value-adding functions, great gains in both cost savings and cycle times were identified. Figure 14 illustrates the financial impact associated with each improvement.

In this particular case, this data was utilized in conjunction with automation and systems development efforts. However, independently of the route chosen, the analysis allowed the development of a performance criteria, very precise outcomes to be achieved as a result of the improvement process (i.e., automation or otherwise). At the present time, this same framework is being used to estimate the impact of Electronic Data Interexchange (EDI) upon the process. Preliminary estimates indicate that by incorporating only half of the supplier base, approximately $302,375/year would be saved (only in the receiving, processing and entering of invoices). At a cost of $30,000 for software, and $22,000 for hardware, communications, etc., the investment on EDI would yield almost a six to one return (year 1). Most importantly, it would greatly contribute to enhancing service quality and overall responsiveness.

A FINAL COMMENT

Quality Function Deployment is in its infancy, and its use is likely to grow. However, be aware of the fact that QFD will never be effective unless it is preceded by accurate market research data. In addition, QFD is a framework, not the solution itself. It is only the beginning of a long journey, and improvement only results from actions, not from analysis alone. Employee involvement to foster the development and implementation of actions to maximize value and eliminate waste will be largely responsible for the outcome. Finally, QFD can not be a "one time deal." Customers change, so does their notion of what is "good." Consequently, organizations must constantly identify and monitor customer value-perceptions. This should trigger a constant renewal of processes and policies. Let us not forget that at the end, customers will always be the ultimate paymasters and heart of all strategies.

REFERENCES

Fortuna, R. M. "Beyond Quality: Taking SPC Upstream." *Quality Progress* (June 1988), pp. 23-28.

Figure 1

A DISCIPLINED WAY OF COMPARING TWO SETS OF ITEMS:

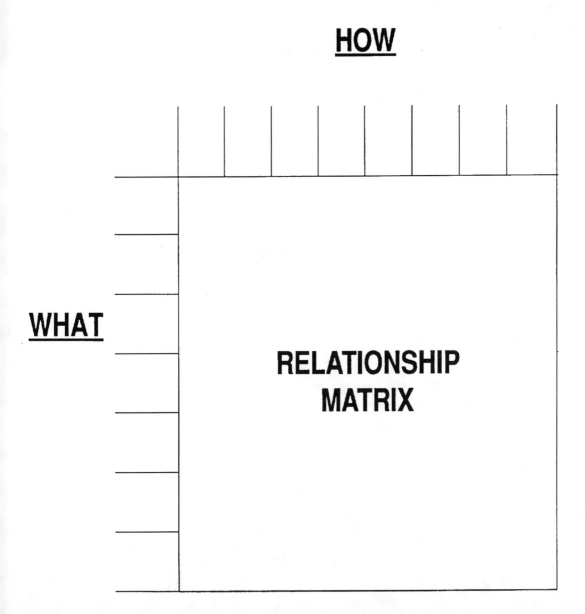

Figure 2

"HOUSE OF QUALITY"

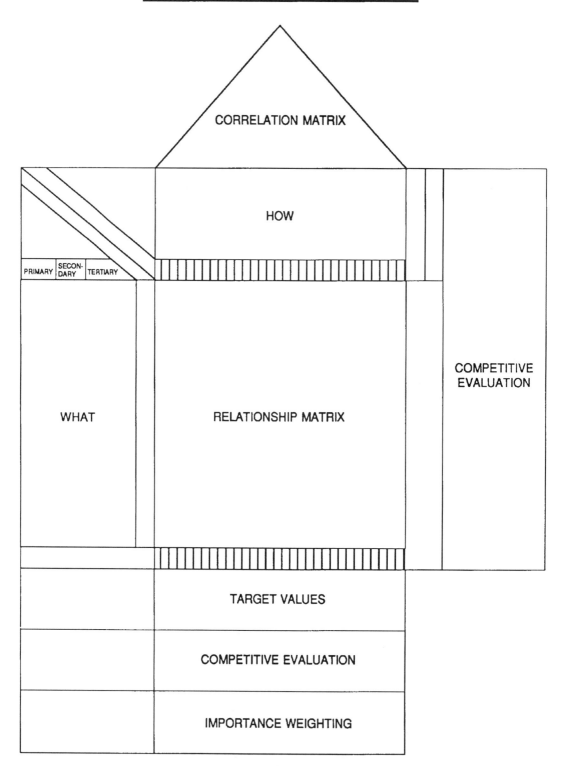

Figure 3

CORRELATING CUSTOMER REQUIREMENTS
TO PRODUCT SPECIFICATIONS

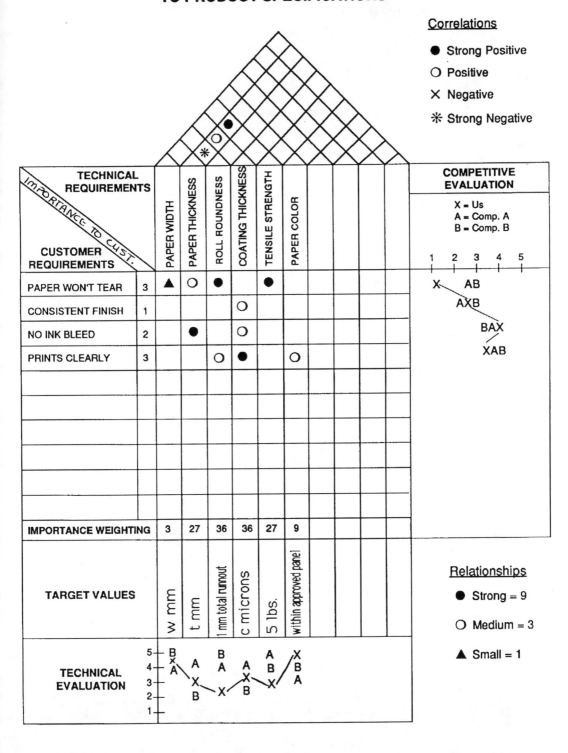

Figure 4 (*)

Deploying the Voice of the Customer

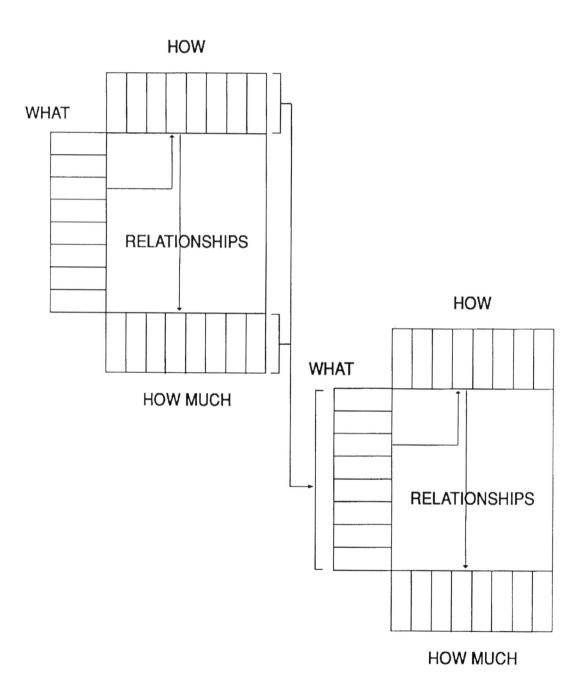

(*) From "Introduction to Quality Function Development," by William E. Eureka, included in "Quality
     Function Deployment," a collection of Presentations and QFD Case Studies. American Supplier
     Institute, 1987

Figure 5

Customer Needs Assessment

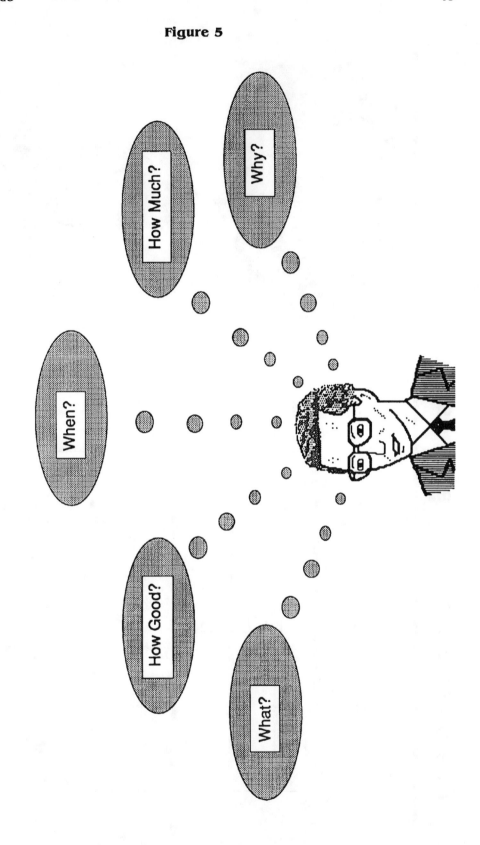

Figure 6

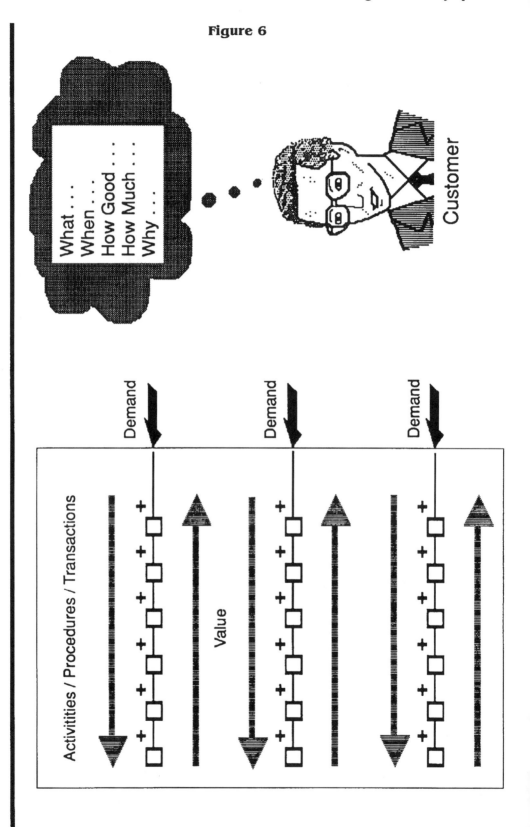

Figure 7

POINTS OF CONTACT AND DEMANDS ON THE ORGANIZATIONS

AREA	DEMANDS		COMMENTS
	WHAT	**HOW**	
Product / Service Planning	• Product Quality • Service Quality • Responsiveness • Price	• Manufacturing • Customer interphase depts. • Distribution	• Good market research data • What do customers buy?
Product / Service Design	• Customer-interphase	• Customer service policies • Information systems (i.e., prices, availability, order status, etc.)	• How does a company deliver value?
	• Manufacturing	• Design / Engineering • Materials	
	• Distribution	• Manufacturing • Warehousing	
Process Planning	• Customer Service Policies	• Processes / Strategies (Customer Service)	• Processes
	• Information Systems	• Systems Design / Hardware / Software	• Activities
	• Design / Engineering	• Drafting / CAD	• Process Value
	• Materials	• Production Department / QA	
	• Warehousing	• Inventory planning / Delivery Schedule	
Process Control Planning	• Customer service processes	• Tasks • Data	• Processing of complaints • Failure prevention
	• Systems Design / Hardware / Software	• Process designs / Activities	• Built-in waste
	• Drafting / CAD	• Design activities	• Built-in waste, idle time
	• Production / QA	• Production planning • Shopfloor level targets	• Target values • Minimize idle time
	• Inventory / Delivery Planning	• Daily schedule planning	• Lead times

Figure 8

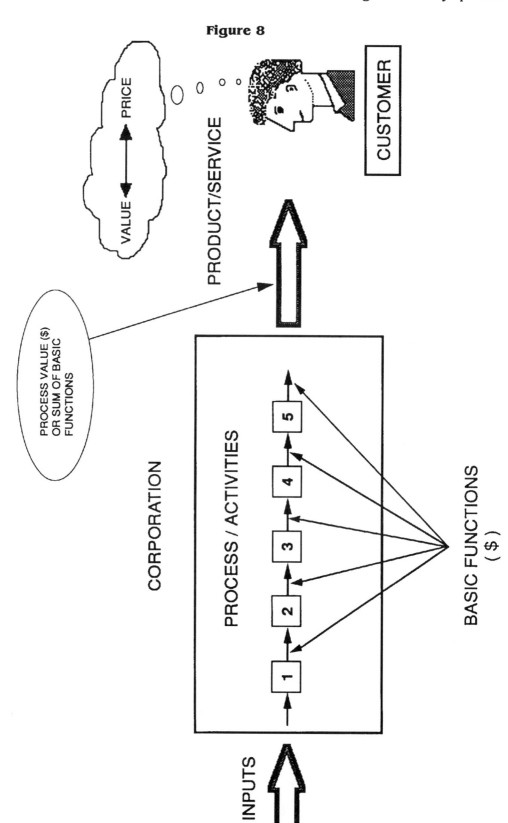

PROCESS VALUE AND CUSTOMER-PERCEIVED VALUE

Figure 9

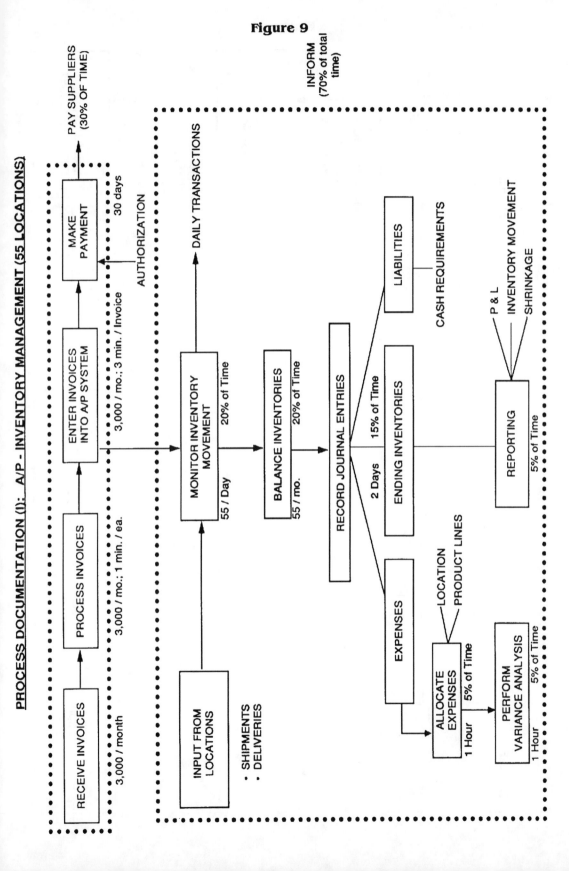

PROCESS DOCUMENTATION (I): A/P - INVENTORY MANAGEMENT (55 LOCATIONS)

Figure 10

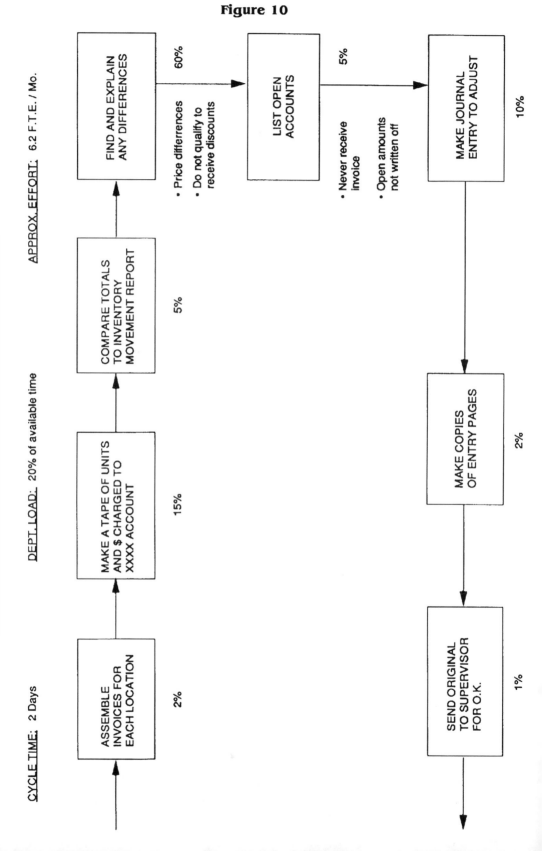

PROCESS ANALYSIS (II): RECORDING JOURNAL ENTRIES

CYCLE TIME: 2 Days

DEPT. LOAD: 20% of available time

APPROX. EFFORT: 6.2 F.T.E. / Mo.

Figure 11

PROCESS VALUE AND PROCESS COSTS: AP - INVENTORY MANAGEMENT (*)

ITEM	ACTIVITY	BASIC FUNCTION	%	COST ($ / YR.)	WORTH ($ / YR.)	COST / WORTH
INVOICES	RECEIVE	SEPARATE INVOICES	30	67,725	20,317	3.3
	PROCESS	CODE INVOICES	50	201,600	100,800	2
	ENTER	ENTER INVOICES	60	335,425	201,600	1.6
	PAYMENT	PREPARE CHECK	30	67,250	20,160	3.3
	TOTAL INVOICES			672,000	342,877	1.95
INVENTORIES	MONITOR	CHECK TRANSACTIONS	5	189,440	9,472	20
	BALANCE	MAKE CORRECTION	10	322,560	32,256	10
	TOTAL INVENTORIES			512,000	41,728	12.26
JOURNAL ENTRIES	EXPENSE	RECORD ENTRY	10	113,066	11,306	10
	INVENTORIES	RECORD ENTRY	10	113,066	11,306	10
	LIABILITES	RECORD ENTRY	10	113,066	11,306	10
OTHER	EXPENSE ALLOCATION	ASSIGN COST	15	19,200	2,880	6.6
	VARIANCE ANALYSIS	CALCULATE DIFFERENCE	50	21,504	10,752	2
	REPORTING	LIST ITEMS	25	21,658	5,414	4
	CASH REQUIREMENTS	CALCULATE REQUIREMENTS	50	14,438	7,219	2
TOTALS				1,600,000	464,948	3.44

(*) all costs ($ / year). White-collar functions supporting 55 locations.

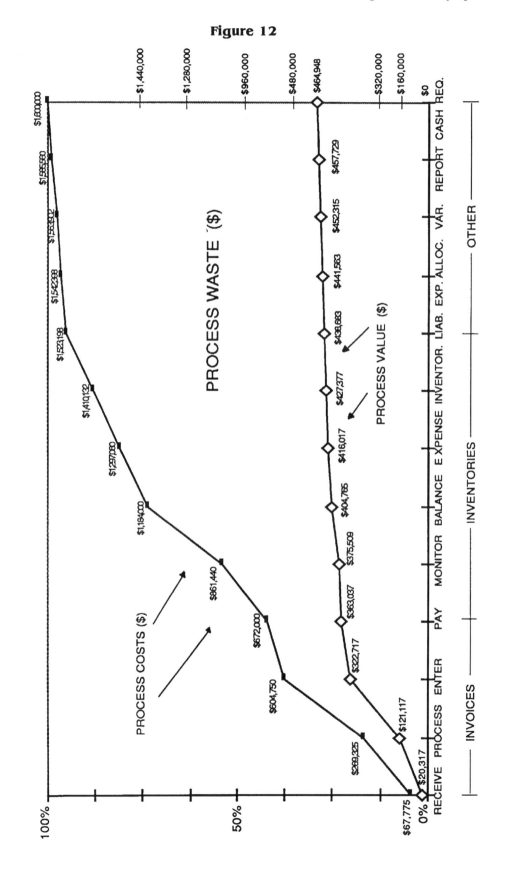

Figure 12

Figure 13

PROCESS IMPROVEMENT FOCUS

AREA	ELIMINATE	SPEED
PROCESSING INVOICES	• Calling purchasing or suppliers if prices do not match ($47,545 / yr) • Checking if units not in G/L System ($54,322)	• Coding
MONITORING INVENTORY TRANSACTIONS	• Assisting locations with balancing problems ($78,798)	• Verifying inventory status reports
BALANCING INVENTORIES	• Calling locations ($12,325 / yr.) • Checking with suppliers ($126,325)	• Making corrections
RECORD JOURNAL ENTRIES	• Finding & explaining differences ($87,111 / yr.)	• Entry

Figure 14

ESTIMATED BENEFITS

1. **PROCESSING OF INVOICES**

   - RECEIVE: $35,326   (FAST SORTING BY MONTH)

   - PROCESSING: $78,353   (FAST / ACCURATE CODING, ELIMINATE WASTE)

   - ENTERING: $107,060   (FAST / ACCURATE ENTRY)

   - PAYMENT: $21,371   (FAST PAYMENT PROCEDURE)

2. **OTHER**

   - VERIFICATION OF STATUS DATA
   - CORRECTIONS
   - RECORDING ENTRIES
   - ELIMINATING WASTE
   - TOTAL SAVINGS = $665,928 / YR

3. **TOTAL POTENTIAL SAVINGS:** $908,041 / YR   (*)

(*) IT AMOUNTS TO ABOUT 80% OF ESTIMATED PROCESS WASTE   ($1,135,052)

CHANGING MANUFACTURING PERFORMANCE MEASUREMENTS

Thomas Vollmann
Professor of Operations
Boston University
Boston, MA 02215

ABSTRACT

Changes in technology and the business environment have rendered traditional accounting methods for measuring performance obsolete. This paper examines the characteristics of traditional performance measures, identifies their weaknesses, and suggests new performance measure alternatives that would improve cost modeling and feedback functions.

Financial information is useful in providing information that can be compared with other organizations. Management decisions based on such cost data take a short term perspective, can harm long range performance, are reactive in nature, and fail to include organizational strategy in the decision making process. Even the current response of altering allocation bases does not explore the root causes of inefficiency or provide incentive to cut production costs and thus will not lead an organization to be globally competitive.

An alternative view of performance measurement views strategy as the essential element guiding decision making and situational factors relating to strategy as primary determinants of specific action. Such system requires continuous feedback of operational activities focusing on long range economic well being. Resources are constantly managed and redefined in order to keep pace with the changing economic environment. If done effectively, an organization not only becomes globally competitive but creates barriers to entry that help ensure its success.

Manufacturing companies are increasingly concerned with how best to evaluate performance in the light of new competitive challenges, the need to roll out more new products faster, the steady decrease in direct labor as a source of value added, and the impact of new manufacturing technologies such as just-in-time. Computer Aided Manufacturing-International (CAMI), other groups, and individual researchers have been working on improved manufacturing performance measures to meet these challenges [Kaplan, 1987]. However, the bulk of this work either attempts to retain a great deal of classical cost accounting (in that cost accounting functions such as individual product costing are preserved), or attempts to provide some monolithic replacement for cost accounting (i.e., the same new measures will be used by most if not all firms).

Both of these approaches seem futile, in that they are keeping us from properly attacking the manufacturing performance measurement problem. It has been said that one gets what one measures: measuring costs may well be the tail wagging the dog. Costs and other financial measures increasingly need to be viewed as a result or "follower" rather than as a cause or "driver" of good or bad manufacturing decisions. Taking this point of view focuses attention on manufacturing strategy as the key element in determining appropriate performance measures. Moreover, this view leads to a contingency model for performance measurement: One needs to measure things that are congruent with the *present* set of perceived strategic objectives. The measurement of performance can and should evolve as strategy evolves. In fact, not changing manufacturing performance measures to support strategy is the *cause* of many present problems of American competitiveness.

The balance of this paper is devoted to presenting a three phase framework for thinking about these issues, and the resultant process of changing manufacturing performance measurements. The first section details many of the shortcomings of traditional cost accounting systems for dealing with contemporary problems in manufacturing. Cost accounting based measurements are increasingly dysfunctional to achieving the manufacturing infrastructure necessary to effectively compete in today's (and more importantly tomorrow's) global marketplace.

The next section of the paper describes the first phase of the three phase framework for changing performance measurements, "Tinkering with Cost Systems." We find many companies modifying their existing systems to redress inadequacies in costing systems, typically by changing overhead allocations. Unfortunately, these efforts are largely misguided since they provide no new values to customers or no new efficiencies in manufacturing. Systems that have been developed to report the financial stewardship position of the firm to outside interests are simply *not* appropriate for internal decision making and control. General Doriot of the Harvard Business School summed it up well in his advice to students: Spend your time making it or selling it—not counting it.

The second phase of the three phase framework for changing manufacturing performance measurement is called "Cutting the Gordian Knot." This phase involves a conscious decision to no longer be constrained by cost accounting in the development of manufacturing performance measures. Financial reporting will not be linked to internal decision making, feedback, and control. Several company examples are described.

The last phase, "Embracing Change," links measurement explicitly with strategy. Manufacturing performance should be contingent upon manufacturing goals. As certain goals are achieved, new ones should be developed. As the marketplace changes, new responses are required. All of this means that performance measures can and should be changed. The companies that position themselves to make these changes faster, with less conflict, will be winners.

SHORTCOMINGS OF TRADITIONAL COST ACCOUNTING SYSTEMS

The Boston University Manufacturing Roundtable has been investigating manufacturing performance measurement for several years. Included in this investigation are the annual surveys of manufacturing companies in North America, Europe and Japan. These surveys examine current practices in manufacturing with regard to perceived manufacturing concerns, priorities, action programs to improve effectiveness, and performance. Also collected are data on cost structures and expected changes in these structures. Several key observations come from the surveys and from follow up discussions with manufacturing executives.

One of the most fundamental observations that can be derived from six years of North American survey information is a growing (but not yet complete) congruence between strategic objectives, priorities, action programs, and performance measures. In the 1982 North American Survey, manufacturing firms were very concerned with quality, and consistent quality has been a top ranked competitive priority in each of the six survey years. However, when we look at the action programs to improve manufacturing effectiveness, in the early years the major effort was on production control systems, not quality improvement programs such as statistical process control. Similarly, yield has only recently come up into the top ten concerns in North America (it was always there in the Japanese Survey). It has taken five or six years for the concerns and action programs to come into congruence with the strategic quality directive.

Unfortunately, there is still not congruence with performance measures. The preponderant performance measures are based on traditional cost accounting with its emphasis on direct labor utilization. Moreover, the problem is exacerbated by the growth in a new strategic objective in many firms: time-based competition. More and more firms are adopting just-in-time approaches. Others see more rapid new product introductions as critical to their long run survival. All of this creates even greater need for new measures that are congruent with the strategic objectives of time and quality.

What we see from the Manufacturing Futures Project and from discussions with manufacturing executives is growing frustration with traditional cost based measures. As quality and time become more important for company survival, measures that are driven by quarterly earnings reports and investment decisions based solely on "cost savings" are ever less relevant to long term company health.

The attainment of corporate objectives is increasingly less determined by the traditional activities of direct laborers working on the shop floor. The factory must routinely execute schedules with high quality and low cost. This is the ante to play in the game. Winning requires more. The routine things need to be done routinely, with more knowledge work done by all employees, both "direct" and "indirect" workers. It is critical to devote more efforts to rolling out new products faster, achieving higher quality, and responding to the vagaries of the marketplace. Achieving these objectives, and having more knowledge work done by everyone in the organization, requires a new set of performance measures (Miller and Vollmann, 1985).

There is increasing dissatisfaction with existing measures of manufacturing performance and the inferences that come from cost accounting systems. The changes that are required to restructure manufacturing operations for global competition are being inhibited by financial measures and controls. Cost accounting systems are built on the premise that direct labor is the major source of value added; this is now obsolete for most companies. Classical systems look upon labor as a resource to be fully utilized in activities that directly create value. The result is a bias toward the accumulation of inventories that can be much more detrimental to overall company performance than approaches that only create products in response to actual customer needs.

Just-in-time (JIT) methods clearly expose the fallacy of building inventories. The concept of "material velocity," where increasing the speed with which materials flow through a factory, is now being viewed as a competitive weapon. Initially used only in repetitive manufacturing environments, JIT is now being applied in non-repetitive manufacturing environments where surges in volume are readily accommodated. For example, a company that makes CAD/CAM/CAE terminals and systems used to have a 15-20 week manufacturing lead time, ran with a steady rate of output, and produced to finished goods. They now produce products in four days, only for firm customer orders, are able to increase or decrease volume levels significantly within a six week window (plus 50%, minus anything) and can increase volume as much as three hundred percent in three months—without adding permanent workers to the employment rolls.

One result of making these changes is that the ways in which performance was evaluated needed to be changed. A casual look at the factory floor in this firm might give the faulty impression that the employees are not being worked hard enough. In fact, the amount of "knowledge work" done by the production workers is quite awesome. Before the JIT system was installed, the company had 86 people in manufacturing planning and control (including MRP planners, purchasing, stock rooms, expediters, etc.). After implementing JIT, the company has only 17 people doing this work. Much of the work of manufacturing planning and control has been eliminated, but some has also been assumed by the direct laborers. What this means is that evaluation of direct laborers with standard cost accounting based methods leads to erroneous conclusions. The objective is to have these people continually assume *more* work usually done by indirect workers, and to determine how to better handle volume surges and new product introductions. None of this is well measured with standard cost accounting methods.

The approach of this company to cost accounting while implementing their JIT systems has been to ignore it. The bottom line results in terms of costs have been quite good (i.e., they *follow*), but no one in the plant uses any cost accounting information for decision making. Cost accounting is viewed as a frustration that has to be put up with. This was not a major problem until the recent appointment of a new comptroller at the divisional level to which this plant reports. He wants to install a classic cost accounting system, measure in great detail, and increase the profitability of this factory. The factory people are not even sure how to respond to this request.

A similar tale of frustration comes from a large computer manufacturer. This firm has found that depending upon how overhead is allocated, they can change the gross margin on a product line by as much as 20 percentage points!

Many such war stories exist. The key points are:

- Managers are frustrated with traditional cost based measurement systems.
- There is a growing irrelevance of "cost" to the newer strategic objectives in manufacturing.
- Cost is an accounting convention subject to wide latitude.
- Cost is best seen as a follower not a driver.
- Managers who put cost and short term financial measures as drivers may be hindering the achievement of long run company health.

PHASE ONE: TINKERING WITH COST SYSTEMS

The first phase of development in changing measures of performance in manufacturing is to tinker or adapt present cost based systems to better reflect reality. The classic response is to attempt new methods for overhead allocation. We have seen several firms decide that since a growing proportion of their product cost is from purchased materials, they should allocate all overhead costs associated with purchases to these items. While this at first seems like a laudatory idea, it has some basic flaws. If, for example, overhead is allocated on the basis of purchasing dollars, that might seem "reasonable." But what about two product lines, one mature and the other new with many engineering changes? Do they both require the same degree of attention from purchasing managers? From engineers? From quality control?

A similar problem occurred at a firm where all of the new products were designed for manufacture on the latest numerical controlled equipment. The older items, especially spare parts, were tooled to be made on the manual machines with longer setup times. Since products were charged at a rate based on shop hours, the old items looked like losers. They took longer to make; moreover, they had to absorb a high overhead because of the depreciation charges from the new equipment—that they did not use! Unfortunately, this firm was talked into a massive study by their accounting firm to more "fairly" allocate overhead. In my opinion, this only increases the overhead by the amount of the study and the new systems required. No efficiencies were gained in manufacturing, no new value was provided to the customers, and the bottom line was reduced. Perhaps the famous line by Bosquet (On the charge of the Light Brigade at Balakava), "It is magnificent, but it is not war," should be altered to describe such work as: It is magnificent, but it is not *competitive* war!

Goldratt and Cox, in their classic: *The Goal* (1984), make the point that an hour of capacity lost at a bottleneck work center is an hour of capacity lost to the entire factory, and therefore extraordinarily costly; while an hour of capacity gained in a non-bottleneck is an illusion that will unnecessarily increase work in process inventory. Cost accounting systems that encourage full utilization of workers in making products are said to be the enemy of productivity. No amount of tinkering with accounting systems will solve this problem.

It is useful to identify three separate activities performed by accounting systems (Nanni, Miller and Vollmann, 1988). The first is financial reporting with the objective of reporting financial health of the enterprise, on a consistent basis with other firms, to outside interests such as shareholders and taxing authorities. The second activity performed by cost accounting systems can be called "cost modeling." Included are studies for pricing, analysis of product line "profitability," study of one-time decisions, and make-buy analysis. The third activity performed by cost accounting systems is feedback and control. This has usually been accomplished with variance reporting and similar approaches.

Separation of accounting into these three categories leads to some useful observations. There will always be a need for category one, and the present methods for reporting to outside interests are probably as good as any (they are not broken, so let's not fix them). It is categories two and three where the problems occur. In both cases the present cost accounting based systems can easily lead to poor decision making. Moreover, tinkering with existing approaches simply will not do the job. As long as the requirement exists that the results of cost accounting will aggregate into the financial reporting (category one), the results will be lacking. Why should a firm consistently perform either cost modeling or feedback/control?

Cost modeling should be done on whatever basis makes the most sense for that problem. There are many long term decisions, such as investments in new technology, that simply should not be based on cost. The real question is what will happen if the firm *does not* make the investment? Similarly, there are short horizon and one-time decisions that are situation dependent.

An illustration of a correct long-term decision was a food company that built two new factories even though the capacity was not required. This was a strategic decision, to put the "bread and butter" production into new, highly-focused factories, so that the main plant could concentrate on new product introductions. The firm believed that the key to long run success was to bring out twice as many new products each year. This would simply not be possible if the present factory also had to push out high volume products. The measures used to evaluate the three factories clearly had to be changed. The older main plant has a completely different competitive charter and has to be measured against new product rollout objectives. The two new plants are evaluated in more traditional ways: cost, quality, schedule performance, and learning curve measures.

A pricing decision example was in a chemical company that produced two families of products from one basic raw material. One family was sold in a highly pure form and commanded higher prices than the other. Cost simply was not the basis for making price decisions in this situation. Moreover, a new use was found for the high purity product, but the same quality requirements were not necessary. However, for this firm it made sense to drive toward constant highest quality production, even if this meant shipping products that exceeded customer requirements. The important "cost" issues in this company were the incremental costs of new products, and the very real, but highly intangible, problems that might occur if customer X found that customer Y was receiving a very similar product at a different price.

Feedback and control, the third activity performed by cost accounting based systems, is even less well fulfilled. For most firms, by the time accounting based variances are determined, it is far too late to do anything about the problems. Moreover, the "cure" may not match the disease. Most accounting measures for feedback and control are a misnomer. They are followers rather than drivers.

We have seen far too many situations where drivers and followers get mixed up to the overall detriment of the company. Examples include building unneeded inventories to utilize capacity and "improve" manufacturing variance measures, restricting capital expenditures to make accounting statements look better, and foregoing important maintenance projects.

The right feedback drivers in many cases have to be in real time and in fine detail. For example, quality control has to be very current, and specific to particular processes. The idea of waiting until a week after month end to determine overall performance is ridiculous for most critical manufacturing variables. It is these variables that need to be controlled. Moreover, if they are held in tight control, the financial variables should work out. If they do *not*, the fault may well be in the financial analysis, not in what is truly important for the company.

A related issue is in evolution of feedback and control reporting. As performance improves new measures and new controls are appropriate. Similarly, when performance slips for some reason, the required level of analysis may easily change. For example, when some products are not up to the desired quality levels, analysis might lead to some area of the factory that needs to be tightly controlled with statistical process control. Later, when the processes can be run with no problems, perhaps the level and degree of control can be relaxed. But, if an unexpected problem occurs, there may be a need for some new highly focused feedback and control mechanisms. The key point here is one of contingency: the conditions dictate the problems and the feedback mechanisms. The approach of cost accounting based systems runs counter to this approach. There, the same set of measures is collected at all times, irrespective of current problem definitions.

The overall result is that the three activities commonly associated with cost accounting are fundamentally incompatible. The problems caused by this incompatibility are growing at a fast pace. The challenges of the marketplace and the new approaches to manufacturing are feeding this growth. International competition demands that companies adopt situation dependent performance measures, ones that are focused on long run economic well being, not short run expediency. The emphasis has to be on managing resources not managing costs; and on deploying resources—particularly human resources.

PHASE TWO: CUTTING THE GORDIAN KNOT

At some point, a few firms have concluded that there simply is no way that a cost accounting system can nor should be used for all three activities. It will always be necessary to report to outside interests on a basis consistent with that of other firms. But there is no reason for consistency in either cost modeling or feedback/control. In both of these activities, the firm should base its actions on whatever best matches its set of strategic objectives.

Saying this and doing it are two different things. Many people will agree with the basic conclusion but will they make the change? Will they cut the Gordian Knot that ties short term financial accounting to decision making in manufacturing? Will they not only adopt new performance measurements, but discard those that are no longer appropriate? What sort of change process needs to occur in a company for this to happen?

One telecommunications firm recently gave up absorption costing for all internal reporting. It was felt that this step would create a new awareness of what was truly important in manufacturing. Accompanying this change in accounting was the participative development of a well defined manufacturing strategy for the company. This strategy clearly defined fundamental objectives for manufacturing in terms of revenue growth, overhead growth, inventory (material velocity), quality, and new product introductions. The strategic objectives led to a definition of what was important to measure, and the abandonment of absorption costing provided a way to unshackle the old measurement constraints.

A similar "knot cutting" has been done by a computer manufacturer. This company has given up standard capital appropriation approaches for a large portion of the budget for improving manufacturing processes. Instead, the firm has made an allocation of funds for what it calls manufacturing research and development. This allocation is to be used to support projects that can be proposed from anywhere in the company. The firm has a standard format for proposals and it has published guidelines for what is and is not included in manufacturing R&D. The proposals are sorted into categories (called "bins"), and each has a bin manager assigned to coordinate and evaluate progress. The size of the appropriation has grown by approximately 50% per year over the past nine years. The aspect most interesting in this example of knot cutting is that the selection process for projects is not based on the usual financially driven criteria. The primary question is the potential for enhancing the products and services offered to customers (Cassidy and Vollmann, 1987).

Another Gordian Knot cutting is a growing tendency among high tech companies to abandon the concept of "direct labor." These firms increasingly view their employees as an asset to be managed and enhanced. They employ people, not direct and indirect. There is work to do, and there will be continual redefinition of who is to do what. The objective is to evolve as quickly as possible (i.e., to *learn*). Employees need to continually increase their skill base and take on new challenges.

The abandonment of direct labor means that accounting necessarily needs to change the basis for many kinds of calculations. Costs of products clearly can not be determined in the same way. This forces people to think less about "cost" and more about the nature of decisions. For example, the focus might shift from some parochial view of product cost to a better understanding of where the knowledge workers of the company are being deployed and whether this is the best use of their time.

A key issue in Gordian Knot cutting is who is going to lead the change process. Our experience indicates that it is critical for the financial function to buy in. Manufacturing can not do it alone, and financial people need to understand that they have to be a part of the solution instead of only being a part of the problem. Without this commitment, the expectations are that the company will not get beyond the "grousing" stage. In the United States, the financial function has significant power in most companies. They can provide the necessary clout to make changes, or they can block the changes.

Another critical component of the Gordian Knot cutting is a well articulated manufacturing strategy. When this is in place, the firm becomes committed to achieving a set of marketplace objectives, and manufacturing is seen as playing a central role in the achievement of those objectives. The result is an agreed upon mission for each manufacturing unit; this more easily leads to a redefinition of what is important, not so important, and what should be measured.

A sense of the need for change in manufacturing performance measures exists in a large number of firms, but there seems to be some lack of understanding of the pervasiveness of the problem and the extent to which existing measurements are impeding progress toward important enterprise objectives. One technique that we have found helpful, whether or not the knot has been cut, is a questionnaire we have designed for assessing performance measurement conflicts.

The questionnaire is typically administered at several levels in a company, and is divided into three sections. The first section asks each respondent to assess the long run importance of about two dozen improvement areas that the firm might undertake to improve effectiveness. A follow on question asks the extent to which present performance measures either inhibit or support improvement in each of the improvement areas. The second section of the questionnaire provides a set of about 40 performance factors which are again assessed in terms of long run importance to the company, and the extent to which they are or are not being measured in the company. The third section of the questionnaire asks the respondent to pick the performance measures from the prior list which he or she feels are used to evaluate his own performance on a daily, weekly, monthly, quarterly, and annual basis.

By administering the questionnaire at several levels in the company it is possible to assess the extent of congruence on several key performance measurement issues. Congruence can be assessed within manufacturing as well as across functional boundaries. In one firm we surveyed people at three levels of authority in manufacturing as well as people in staff areas of manufacturing and other functions. The results provided a diagnosis of the gaps between important improvements and current measurement support, and the congruence (or lack thereof) of important areas for improvement. For example, in one case we found a great deal of disparity about both the importance and measurement support for computer integrated manufacturing (CIM). This company had not done its homework on defining where CIM fits into their needs. Another common result is for an individual to rate far too many things as being of greatest importance. If everything is equally important, then there is not an overriding sense of priority for coordinating improvement efforts.

The questionnaire should not be regarded as a one time event. As strategy changes, the questionnaire can be readministered to see the extent to which goals have changed, and the present degree of congruence among concerns, action programs, and measurements.

Gordian Knot cutting is essential to making real changes in manufacturing performance measurements. The company needs to understand that tinkering with existing cost accounting based systems simply will not yield the desired results. It is necessary to untie the shackles that are inherent in these systems. Situation dependent measures need to be established, and evolved as situations (and strategies) change.

PHASE THREE: EMBRACING CHANGE

The final phase of development in changing manufacturing performance measures is to consider the process of performance measurement as an integral facet of manufacturing strategy. That is, if strategic goals are to be developed and reached, it is critical to develop performance measures that are supportive of these objectives, and to get rid of any measures that are counterproductive. Moreover, as goals are achieved and new ones formulated, performance measurement should similarly evolve. In fact, an interesting chicken and egg question arises from this line of discussion. Perhaps the *first* thing one should do to implement a change in strategic direction is to consider what changes in performance measures might be most conducive to achieving the change.

Changes in performance measures need to be seen as both a top down and bottom up exercise. The top management and strategy should determine the overall direction, but the ways in which those goals are to be achieved, and the best ways to support them in particular organizational units are situation dependent. A top down objective to reduce the time to introduce new products might lead to quite different programs in particular parts of the company. In one firm, achieving this objective required a combining of design and industrial engineering; a new culture of cooperation was required, and new career progression steps were initiated. In another firm, one emphasis in speeding new product introduction was on bringing new products into production with as few subsequent modifications as possible. This required getting rid of a performance measure for cost reductions. There had previously been a cost reduction budget for manufacturing engineering. Every year they were evaluated on how much money was saved through redesign of the products and processes to save costs. Unfortunately, this led to products being poorly designed in the first place. The new emphasis was on achieving "mature" cost, which was defined as 110 percent of final cost, within six months of the start of manufacturing. Instead of a cost savings being treated as something good, it was now considered to be a "design error" if the technology in the improvement had been available at the time of product design.

This last example shows both a new set of performance measures and the elimination of an old set. Both of these are important in phase three. The overall goal is to speed the learning process, to achieve new improvements in manufacturing effectiveness faster. This means that a performance measure will only have importance as long as it drives manufacturing to achieve new levels of effectiveness. In theory, any measure should be discarded when another will do a better job. The problem is that an organization can only absorb directional change at a limited rate. Establishing the means for working both top down and bottom up can help in speeding the evolution in performance measures.

Elimination of performance measures can be as useful as adding them. In this day of high powered computers, it is tempting to add more and more measures, without discarding any. Thus, some people will suggest that traditional cost measures be kept and that new measures be added. A good lesson can be learned from the service industries. At the fast food chains such as McDonalds, the primary measure is simply volume per time period expressed in dollars. The restaurants work with such a high material velocity that there is little point keeping track of any inventories. The output is in essence the input as well. As manufacturing firms move to JIT the same ideas are true. If quality can be guaranteed and high material velocity achieved, many standard measures can be eliminated. Manufacturing's job is to take any order and fill it in a lead time that appears as if the products were held in stock, when in fact little or no finished goods are held.

Phase three is called "embracing change" because the expectation is that performance measures should be continually improved. Manufacturing excellence is the objective, and the definition of excellence is clearly contingent upon what is *now* being achieved, the challenges from competitors, bench marking against the best in any activity in any company—not just a competitor, new technologies and their associated opportunities, and new ideas for enhancing the products and services provided to our customers.

Bench marking is an important activity in many companies that are changing performance measures. The process needs to be done carefully, however. There is a temptation to benchmark against existing competitors. This is dangerous. Just because one's present competitors are bad doesn't mean that this happy state will continue. In this day of international competition, a new competitor can come from nowhere, in a big hurry. The questions for bench marking should be based on whoever is the best—in any industry, in any country. Achieving this degree of excellence should be seen as a strategic objective; as a barrier to entry for competition.

An example is a capital goods manufacturer that was the market leader. The major question at a manufacturing strategy session was how to keep that leadership, how to construct greater barriers to entry, and what kind of preemptive manufacturing moves should be made. In this industry, 12-20 week deliveries were the norm. The company decided that if it could provide three-four week delivery on 80 percent of the product line, it would be exceedingly difficult for anyone to compete with them. This goal became the driver for manufacturing changes, both in the manufacturing processes and in the appropriate measurements for manufacturing.

This example illustrates that manufacturing excellence will be achieved through a series of routine actions and projects, directed by an overriding manufacturing strategy, and a consistent set of performance measures. The measurement of performance in day to day activities and one time projects needs to be consistent with continually refined, situation dependent, definitions of excellence.

SUMMARY AND CONCLUSIONS

This paper has presented a three phase change process for improving the measurement of manufacturing performance. It is not enough for the performance measurements to be neutral, i.e., to not impede. One gets what one measures, and it is essential to formulate performance measurements that encourage rapid learning. Manufacturing companies need to get better—and to do so faster. This requires that every person in the organization use his or her abilities to a continually higher degree. Everyone needs to work smarter, and the measurement system needs to encourage this development. In order to do so, it will be increasingly necessary to embrace change.

REFERENCES

Cassidy, F. D., Vollmann, T. E., "Enhancing Manufacturing Processes" (Working Paper, Boston University Manufacturing Roundtable, 1987).

Goldratt, E. M., Cox, J., *The Goal* (North River Press, 1984).

Kaplan, R., "Yesterday's Accounting Undermines Production," *Harvard Business Review* (July-August, 1987).

Miller, J. G., Vollmann, T. E., "The Hidden Factory," *Harvard Business Review* (September-October, 1985).

Nanni, A. J., Miller, J. G., Vollmann, T. E., "What Shall We Account For?" *Management Accounting* (January, 1988).

ACTIVITY MANAGEMENT AND PERFORMANCE MEASUREMENT IN A SERVICE ORGANIZATION

H. Thomas Johnson
Retzlaff Professor of Cost Management
Portland State University
Portland, OR 97207

Gail J. Fults
Professor of Accounting
Humboldt State University
Arcata, CA 95521

Paul Jackson
Security Pacific Automation
275 S. Valencia Ave (B2-31)
Brea, CA 92621

ABSTRACT

This paper is a case study of a service organization which undertook an Activity Management Analysis in an effort to control costs because it deemed traditional cost management techniques inadequate in its radically changing competitive environment.

Traditional cost management techniques do not focus on non-financial issues like quality, service and flexibility, and often encourage functional separation of work and high volume production which can raise total cost and not necessarily add value to the product.

Activity management examines an organization's activities, who does them, why they are done, and whether they are essential to the strategic objectives of the organization. Keeping in mind both the organizational strategies and the nature of the work that is being done, an activity flow design is created. Performance measures are designed to identify and encourage the flow of value through the system. Once an organization meets its activity based performance goals, it must then strive to reduce its time and resource usage while maintaining such levels. This is what leads to global competitiveness.

Two recent changes in the American telecommunication industry's competitive environment have prompted managers of former Bell operating companies to evaluate the usefulness of long-standing approaches to managing costs. First, divestiture from AT&T in 1984 has made it necessary for these managers to address the demands of capital markets for the first time. Second, technological developments since the 1970s have created "bypass" opportunities for non-regulated competitors to serve customers who once were considered the exclusive preserve of regulated providers.

These changes shift the identity of the operating company's customer from the regulator to the end user of telecommunications services. To be competitive and profitable, therefore, the operating company more than ever requires flexible, adaptive, cost-conscious management. Whereas profitability once hinged on managements' ability to negotiate cost-plus tariff schedules, it now results increasingly from their ability to control costs and manage product-line strategies.

In an effort to improve Pacific Bell's efficiency and profitability, top management in 1987 targeted a company-wide 25 percent reduction in overhead expense over three years. This need to reduce costs stimulated managers in the customer billing unit of Pacific Bell to experiment with a new approach to cost management. Their existing control system emphasized financial budgets and a short list of non-financial service targets. However, managers in Pacific Bell's customer billing organization were increasingly uneasy about their ability to use traditional financial budgets and service targets to control costs. They elected to experiment with a new approach to cost management that emphasizes eliminating waste from resource-consuming activities.

The locus of their experiment was a project carried out between June 1987 and February 1988 in the company's Cash Management Center in Van Nuys, California.. The Van Nuys Center in 1987 employed about 112 persons, including managers, and it processed about four million customer payments a month. It was selected as a project site because top management believed it offered a robust setting to test the efficacy of new cost management ideas. The Van Nuys Center has fairly self-contained operations, with minimum spill-over into other parts of the company, and it has a long-standing reputation for efficient performance. This paper describes the Van Nuys project and the new approach taken there to managing both overhead cost and production activities.

THE VAN NUYS COST MANAGEMENT PROJECT

The team that directed the Van Nuys cost management project consisted of the three authors of this paper, one of whom was employed in Pacific Bell's customer billing organization at the time, and two persons from the management consulting unit of Arthur Andersen & Co. At all times during the project the team worked closely with employees and managers of Pacific Bell's customer billing unit, especially those in the Van Nuys Cash Management Center.

The project team began by mapping the work people do to accomplish the Cash Management Center's three main objectives—post payments to customers' accounts, deposit payments in the bank, and reconcile the amounts posted and deposited. The purpose of this mapping was not to portray the flow of work per se. The company had already done that in flow charts showing basic activities the center's personnel performed to get payments from the post office to the bank. Instead, the team analyzed work in the center in order to discover how much of the work now being done would not be required if all payments were processed in the most expeditious way.

The team assumed that the most expeditious handling of payments would be to pass every payment through its necessary stages in an uninterrupted continuous flow, never handling a payment more than once in any process. The reality was quite different. Payments were processed through stages in large batches, not in a continuous flow. Much of the work people did consisted of rework—that is, repeating steps performed previously, in some cases several times.

The project team made an important discovery about this rework when they mapped the center's activities by the different types of customer payments that arrived in the mail. They discovered that the extent of reworking varied considerably among these different types of payment. To understand the import of this discovery, let's define the different types of customer payments and the different steps through which payments are processed:

1. Types of payments.
   - **singles (about 66%):** payments from one customer with one stub and one check, received in the company's return mail envelope
   - **multiples (about 23%):** payments from one customer with more than one stub (signifying more than one phone line) and/or more than one check per envelope
   - **white mail (about 11%):** payments mailed in a non-company envelope
   - **problem mail (< 1%; some included in the first three types):** torn stubs or envelopes, checks stapled to stubs, Scotch tape on bills, other forms of damaged mail

2. Steps to process payments.
   - **sorting:** The sorting department separates singles, multiples, white, wrongly addressed, and problem mail. They also record estimated counts of checks received each hour. Sorted mail goes in standard-sized trays to the opening department.
   - **opening:** Personnel in the opening department open singles by machine and multiples and white mail by hand. They align payment stubs with checks and send them either directly to the IBM 3895 computer processing unit or to one of the manual processing units. They also sort out and batch for later processing any stapled, taped, torn, or disputed items that require manual attention before going to the 3895 room.
   - **3895 prep:** To prepare complex and error-ridden payments for processing on the IBM 3895 machines, personnel perform myriad tasks such as prebalancing multiple payments and repairing damaged mail.
   - **manual processing:** Several sub-departments manually process items that can not go through the 3895 machines. This work includes things such as researching on-line customer files to correct billing errors or customer claims, handling credit card payments, handling payments from agencies, and so forth. Personnel in this department also post customer accounts directly through an on-line database and forward checks directly to the department that prepares bank deposits.
   - **3895 processing:** The IBM 3895 (four in service) accesses customer record files, reads both stubs and checks, MICR encodes killed checks and sorts them by bank. Because a 3895 is so fast—17,000 items per hour—managers push large batches into it as fast as possible, even when many items in a batch are not ready to be read by the machine. Checks and stubs rejected by the machine are manually keyed up to three different times before finally being shunted over to manual processing.
   - **deposit prep:** Killed checks from the 3895 and checks received from manual processing, all sorted by bank, are balanced and prepared for bank deposit.

Figure 1 shows the steps that cash management personnel perform to process the four different types of payments received each day. The line at the top of Figure 1 shows the "fast track" that payments would follow if each day's mail contained only error-free singles payments. On average, about one-third of each day's payments follow the track for error-free singles. The lines below the top one show most (but not all) of the additional steps cash management personnel perform to process more complex payments and errors.

Figure 1
Flow of Work in Van Nuys Cash Management Center in Early 1987

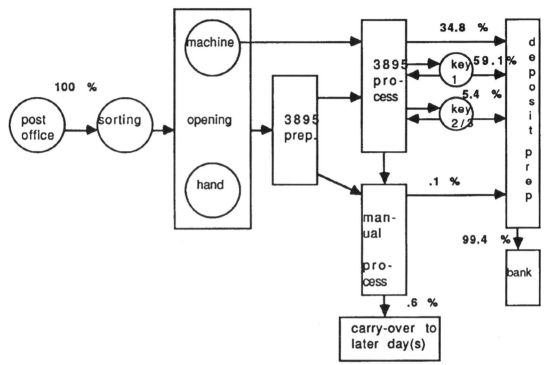

Note: percentages represent average amount of day's payments in May 1987.

Whereas the singles payments flowing along the top line receive only "one-touch" at each process step, most other payments are handled several times in the same process (especially in 3895 prep and 3895 processing) before being posted and deposited. The cost management project team considered it axiomatic that much of this redundant handling could be eliminated if all payments received only one-touch in each process step. While different types of payments might require different processing to be posted and deposited, there is no reason for repeating the same step in order to process any one payment.

After mapping the work currently performed to process payments, the project team then set out to explain why the center processes payments as it does and how it might save resources by processing payments differently.

In interviews, the manager of the center and her supervisors attributed their approach to processing payments to two measures the company uses to evaluate the manager's performance. One measure is the percentage of each day's payments banked the same day and the second is the number of invalid disconnects (customers cut off for delinquent payment when slow processing of the payment caused it not to be posted on a timely basis). For years the company has judged the cash management center's performance to be adequate if 97 percent or better of each day's payments are banked the same day and there are no invalid disconnects. It is also important for the center to meet annual dollar and manpower budgets, but the two "service" targets take priority over budget targets.

Her desire to look well by these service targets—percent of day's payments banked and invalid disconnects—had prompted the manager of the center many years ago to follow two principles in organizing the center's work. First, she pushes payments through the system as fast as possible each day. The reason for this is that the size or status of payment in any particular envelope is unknown until it is opened. Hence, pushing all payments into processing as fast as possible minimizes the chance of carrying big-dollar payments over to the next day, and it reduces the chance of posting late payments after the day received. Moreover, she copes with the uncertain variability of daily payment volumes by starting mail pickup and sorting early each day, about 2:00 a.m., and planning for overtime by counting the day's volume as soon as possible after the mail arrives.

Her second principle is to organize the center's work into departments that specialize by functional activity—that is, sorting mail, opening mail, 3895 processing, and so forth. Each department is encouraged to process its special work as rapidly as possible. The purpose here is to push each morning's "wave" of mail as quickly as possible toward the high-speed automated IBM 3895 machines where checks are automatically posted and processed for deposit (that is, "killed"). Although the 3895s can not process all payments in the form in which they arrive (checks may not agree with stubs, checks or stubs may not be readable, and so forth), their high speed (17,000 checks per hour) causes the center to use them as a screen to sort out those payments that are readable by the 3895s and kill for deposit on first pass. Payments that don't make it on first pass are batched for manual keying and other processing before being passed through the 3895 again.

A source of much of the rework done in the center each day is this strategy of pushing payments into 3895 processing whether ready or not. Obviously any payment that does not kill on its first pass through the 3895 machines is a payment that will be processed for posting and deposit *at least* twice, and often more than twice. In fact, the usual next step with payments rejected on first pass by the 3895 is for operators to manually key the amount on each stub for MICR check-encoding, and then pass the payments through the 3895s again. Most payments make it on the second pass. For those that don't (due to keying errors and so forth), manual keying is done up to two more times. Payments that still don't make it after four passes through the 3895s are then processed and entered into customer accounts manually, through an on-line terminal. Batches of payments needing this attention are often carried over from one day to the next.

Another source of redundant work (and excess movement of work) is the practice of flowing work in large batches through functionally-specialized departments. A floor map of the center's departments in Figure 2 shows a classic functional layout so well known to managers in factories. By separating the functions required to process a payment, from sorting mail to depositing checks, this layout increases the distance work must move and it prevents workers from detecting their own (and customers') errors and correcting them when they first occur. When performing repetitively the same step on large batches of payments, a person often does not realize if they have done something wrong (e.g., misalign a check and stub or miskey an amount). They may, therefore, commit the same error again and again, causing more work later in another department.

To estimate the amount of excess work caused by the center's approach to organizing its activities, the project team measured the time devoted to each type of payment in each separate activity. Data to make these estimates were gathered in two steps. First, the team estimated the average time personnel devote daily to each activity by interviewing people at their work and by gathering information from supervisors. Then they collected data to estimate the average daily number of

Figure 2

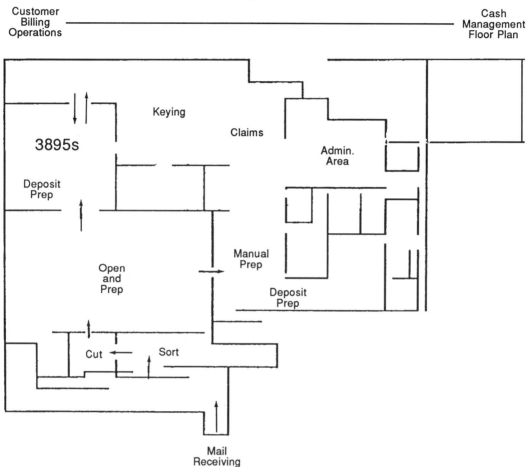

payments processed in each activity. From this information they could estimate roughly the time spent on each type of payment in every activity, and by summing these amounts calculate the total time spent to process each type of payment.

The project team took the added step of estimating the cost to process each payment. The cost data can be used to evaluate the strategic profit and loss consequences of decisions to discontinue certain types of payments or to subcontract payment processing to outside parties. However, the team was able to evaluate the center's operations and to make recommendations for changing those operations without using this cost data. Using their estimates of the time required to process different types of payments, the team analyzed the center's operations and reached the following conclusions:

- About one-fourth of the center's manhours are spent working on about one-tenth of one percent of the payments processed each day. Much labor is devoted to multiple handling of items and to handling errors.

- More than one-third of labor time goes into preparing payments for the 3895 machines that don't "kill" on their first pass. A second pass takes nearly double and a third pass almost quintuple the time to process as a payment that kills on the first pass. Obviously much of this work would not be necessary if a way could be found to post and deposit such payments without handling them more than once.

- Batch-processing damaged mail, as the center now does it, is very time-consuming. Much of the added time is due to multiple handling of payments with staples, tape, rips, etc. Instead of correcting damaged items when the mail is first opened, they are set aside and batched for repair and processing after the day's first rush of payment processing has passed. Correcting the error when it is first apparent saves time and reduces rework.

- Credit card payments take nearly *100 times* as long to process as error-free singles. Either a new way should be found to process these payments or the marketing department should be advised of the costliness of credit card payments and this payment policy reevaluated.

THE PROJECT TEAM'S RECOMMENDATIONS FOR CHANGE

By viewing the center's work in terms of activities required to process diverse types of payments, the project team discovered just how much work in the center represents waste—that is, nonvalue activity—that seems "necessary" because of the way the center's operations are organized. Waste includes all repeat handling of payments, whether caused by the policy of batching all "problems" for later attention or by the policy of forcing virtually all payments through 3895 processing.

To reduce nonvalue work the project team recommended correcting all problems and making all preparations for deposit at the start, as soon as mail is opened. The 3895 machines would not be given anything they can't handle on the first pass and damaged mail would be repaired as soon as it is opened. This would reduce the number of times payments are handled.

The redesigned system depicted briefly in Figure 3 shows error-free singles being processed just as they are now. After being sorted, machine-opened, and stubs and checks properly aligned, these payments would be put into the 3895s for posting and deposit preparation. Virtually all other payments would be handled in two "multi-function work cells," featuring workstations capable of check reading and payment posting. One work cell would process white mail, the other would

Figure 3
**Work Flow for Van Nuys Cash Management Center Proposed by Cost
Management Project Team**

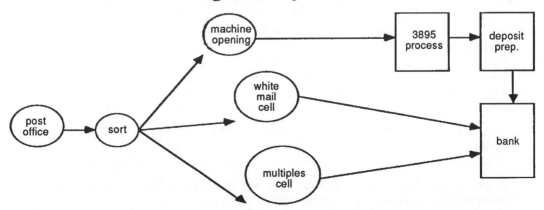

process multiples. The objective in each cell would be to have one person, or one group of persons, handle all steps at once including opening the mail, posting, and preparing checks for deposit. Payments would be processed in "lot sizes of one," with each payment completely handled before another is put into process. This would eliminate double-handling to prebalance multiples, open white mail, remove staples, repair torn and unreadable stubs, research minor discrepancies or errors, etc. It would provide a continuous flow of paperwork from the mail room to the bank, not just for error-free singles but for virtually all payments received each day.

Continuous flow would be achieved, of course, by separating work into different streams that flow according to the rates at which diverse payments can be processed—the rate inherent in the nature of the work itself. Singles would flow in the fastest stream, using the IBM 3895 machines. Multiples and white mail would each flow in work cells at different rates, undoubtedly slower than singles. However, all payments would pass through a continuous flow of "one-touch" processing. That reduces the overall time spent in getting each day's mail into the bank because it eliminates nonvalue duplicate handling. Indeed, about one-third to one-half of all work performed in most offices or factories adds nothing to accomplishing the organization's ultimate objectives. The Van Nuys Cash Management Center's multiple handling of payments is no exception to that general observation.

The project team also recommended a change that the manager of the center herself had proposed several times in the past. This was to print bar code messages on the return stub of all customer bills, showing the amount of payment billed (at least in a range) and the due date. Bar coding would not change the way payments are processed. But by enabling employees to sort unopened envelopes with large payments or critical due dates, bar coding would alleviate much of the pressure to force-feed each day's mail through the system as fast as possible.

NEW PERFORMANCE MEASURES FOR NEW OPERATIONS

Traditional performance measures often stand in the way of putting new competitive practices to work. The project team's recommendation that the Van Nuys Center adopt continuous flow waste-free processing, as with front-end multifunction work cells, promises to increase productivity and improve performance in the center. However, modifications such as work cells represent a radical change in the way the center has traditionally organized its activities. The company's long-standing practice of measuring the percentage of each day's cash banked the same day profoundly reinforces the existing way of running operations. Were that measure to remain in place it could act as an anchor inhibiting the Cash Management Center manager from adopting new practices. New measures are needed to reinforce the one-touch continuous flow mentality that underlies a new development such as the work cell concept.

The project team recommended several measures as being appropriate to a competitive cash management shop in the global economy. Viewing rapid and continuous flow of value-adding work as the hallmarks of competitiveness in the global economy, they proposed measures that would do two things:

- motivate actions aimed at removing impediments to continuous flow of value-adding work;
- confirm success at achieving continuous flow of value-adding work.

They needed measures appropriate both for processing singles payments on the 3895s and for processing non-singles in work cells.

For payments processed on the IBM 3895 machines, the project team proposed tracking the following data: number of payments; number of checks killed for deposit; number of passes payments make through the machines; number of checks killed on first pass through the machines. These data will form two ratios that monitor success at achieving "one-touch" continuous flow of work through the 3895 machines. One ratio is the number of passes to the number of checks killed. The other ratio is the number of checks killed to the number killed on first pass. Both ratios are global indicators of how closely the payments processed on the 3895 machines come to one-touch continuous flow. Under perfect conditions both ratios would equal one. The goal of continuous improvement is to push these ratios closer to one every period.

Measuring payments processed in the work cells will not be so essential. Work done in the cells will be highly visible. The work flow will be designed for one-touch processing. Problems will be easy to spot when they occur. But it is useful to track cell downtime, carryovers of payments not processed on the day received, invalid disconnects, and other errors. Those numbers should begin to approach zero as soon as the cells approach a continuous flow.

Several measures of nonvalue activity are needed for the center as a whole. Among these are: elapsed time to process each class of payment; distances payments move in being processed; space occupied by the center's activities; errors committed in processing payments; and number of payments not deposited on day received. The operating goal is to embark on programs that continually drive these measures to zero.

One global measure to evaluate overall operating performance in the center is the ratio of checks killed per hour to the number of employees in the center. A long-run increase in this ratio indicates that the Cash Management Center is reducing multiple handling of payments, economizing on time, and economizing on number of employees. When coupled with carryover data, this ratio resembles ROI or revenue per employee—an overall measure of efficiency and effectiveness. When the center reaches world-class status, 100 percent of all payments will regularly be banked on the first day. Then the goal is to cut elapsed time and manpower required to deposit and post 100 percent of all payments in a day.

PROJECT RESULTS: ACTUAL AND PROJECTED

The Van Nuys Cash Management Center received the project team's recommendations in August 1987. By February 1988, when the team met in Van Nuys for the last time, the manager of the center had implemented several of the team's recommended changes, including a system to read bar codes on payment stubs and on-line repair of damaged mail at the time it is opened. No effort had been made to implement work cells. Therefore, we can report some actual results of changes in existing procedures; but we can report only the proforma results of adopting work cells as set out by the project team in its August 1987 report.

Adopting front-end work cells to process virtually all non-singles payments would lead, theoretically, to fewer steps (less repeat handling and rework), less distance traveled and less waiting between steps (due to elimination of batch flows from department to department), and shorter elapsed time to process a payment. The overall result of these differences would be less manpower needed to get the job done and less need for overtime (due to a more even rate of payment processing). In return, there would be a need to invest funds in work stations and to change the layout of the facility.

The consultants on the team from Arthur Andersen & Co. prepared a rough estimate of the savings by activity for each type of payment to be processed on front-end work cells. Significant savings would occur in all stages of payment processing, but the lion's share would come from elimination of rework, especially in preparing payments for posting and deposit. Without disclosing confidential dollar amounts, the total annual gross savings amounts to approximately one-quarter of the Cash Management Center's present budget. The estimated payback period for all capital outlays, including floor relayouts and training, is about one year.

Included in that projection was an outlay for a bar code reader/sorter. That piece of equipment the center did in fact purchase and put into operation in late 1987. Along with bar coding payment stubs, the center also made the following changes at that time: people began to repair damage (e.g., remove staples, tape torn checks, remove correspondence, and so forth) at the time they opened mail, rather than off-loading and batching it for later processing; the center picked up mail from the post office at the start of the day, instead of several times in the night before; they adopted even-level processing of payments at a rate of about 165,000 per day. The net result of these changes by the end of January 1988 was a total elimination of overtime work, the elimination of a night shift (formerly used to get payments ready for fast processing by 6:00 a.m. each day), and a reduction in force of seven nonmanagerial personnel and four management personnel (out of a total of 112 employees).

The manager of the center still reports "percentage of day's payments banked on the same day" and "invalid disconnects." But the less hectic pace made possible by bar code equipment now permits her to achieve both targets with less effort and less consumption of resources than ever. Major improvements in productivity can still be made, however, by taking steps someday to off-load all non-singles payments from the 3895 machines.

Further confirmation of the productivity improvements one can achieve by eliminating waste from activities is provided by a project completed in late 1988 in another part of Pacific Bell's customer billing organization, the Bill Print Center in Anaheim. The member of the Van Nuys project team from Pacific Bell, Jackson, was assigned to the Anaheim Bill Print Center in 1988. He immediately engaged all personnel in that center in a search for waste activity. They analyzed the flow of activity according to the different types of print work done in the center and discovered that unevenness in work flows was a major source of waste activity. For several months in the middle of 1988 they pursued projects aimed at reducing unevenness. After five months their efforts led to the elimination of much rework and a reduction in average lot sizes and inventory buffers. These changes at the end of five months reduced manpower needs by twelve persons.

CONCLUSION

The ultimate goal of any business is to be competitive. A competitive business provides value to the customer and thereby maintains or increases its market share. Companies striving to be competitive in the global economy must know if current operating activities—the resource-consuming work people do in a business—are value-adding or nonvalue-adding. Clearly it is desirable to eliminate all nonvalue-adding activity, i.e., waste. To know if their operating activities are helping the business be competitive, managers need information that identifies generators of waste.

Performance measurement systems used by most businesses in the past fifty years do not provide companies striving to be competitive today with relevant information about value. Indeed, standard cost targets and budget variances reported by accounting-based measurement systems encourage behavior that is inimical to competitiveness today. By focusing exclusively on accounting costs these systems fail to identify and measure sources of customer value such as dependability, flexibility, and quality. Moreover, they embody the belief that profitability comes from achieving low costs through economies of large scale and large throughput. Traditional accounting-based cost variances encourage managers to believe that costs are minimized and profits maximized by keeping people and machines busy, no matter what.

Today's profitable and competitive companies achieve low cost and flexibility by achieving continuous flow of value-adding work. They are not fooled by standard-cost variance "efficiencies" that come with uneven work flows and large, inflexible batches. They recognize the need, therefore, to replace yesterday's performance measures with new measures that encourage elimination of waste and continuousness of flow.

Pacific Bell's customer billing organization began to address this challenge in its 1987 cost management experiment at the Van Nuys Cash Management Center. The project indicates that companies desiring to improve their competitiveness and productivity have much to gain by turning away from traditional budget-based cost controls and examining, instead, the assumptions underlying the way work is organized. Competitiveness in the global economy requires a changed outlook, and changed performance measures to support that outlook. Traditional cost control systems fail to address the benefits of continuous flow, small lots, fast changeover, and elimination of waste in activities. These ideas have gained a strong hold in the minds of manufacturers. Pacific Bell's cost management project shows how managers in service establishments also have much to gain from these ideas.

THE IMPACT OF VARIATION ON OPERATING SYSTEM PERFORMANCE

James M. Reeve
Associate Professor of Accounting
University of Tennessee
Knoxville, TN 37996

ABSTRACT

Variation in production processes or at systems levels that affect production levels results in inefficiencies and cause increased production costs. Statistical Process Control is a technique that will identify and help correct variation in production processes.

Traditional cost control techniques use variances to identify problem areas in production. Variance analysis aggregates errors so that they are difficult to trace back to production, is reported long after the events causing the problems have occurred, encourage production goals that just meet a standard and do not promote excellence, and result in retrospective decision making.

Statistical process control instead identifies strategic production targets, measures the state of control existing in the production process by measuring its variation, and leads to determination of causes of variation. Operational changes that will eliminate production variation can then be implemented. Constant measurement of variation will encourage uniform production, increase throughput and decrease inventory requirements.

In the world class firm the responsibility of management is to create and continuously improve systems that produce products and services that are valued by internal and external customers. These systems will be cross-functional in nature, owned by the manager, have both measurable inputs and outputs, and will be demonstrably value added. The emphasis upon the personal ownership of these systems is paramount to the work of creating effective and responsive enterprises. The job of management will move away from fighting the fires of system exceptions towards an active responsibility to define and improve systems.

The methodology of improving systems requires a time ordered record of system outputs. This is a much different measurement requirement than found in traditional control systems. The traditional management control system relies upon aggregate measures of the *level* of performance for a specified period of time. Such measurements are incomplete in a number of ways. For example, data aggregation over a period of time causes the loss of important time ordered information that could reveal causal factors underlying system behavior. Additionally, traditional control approaches emphasize the *level* of output attainment, but systematically ignore the issue of system variation. System *variation* is paramount to system control, improvement, and cost reduction. The use of statistical control charts overcome these two objections and are becoming part of an expanded set of tools for managing and controlling systems. This paper will address these issues in detail.

THE DEFICIENCIES OF ENGINEERED COST CENTERS

The typical control model employed by most enterprises is illustrated in Figure 1. Resources are input to a process for transformation to a measurable output. Control is achieved through the use of engineered standards that are developed for the transforming equation between inputs and outputs. The level of control for determining efficiency of the transformation can be as narrowly defined as an individual or a machine. Management reports will disclose the gross variance over a period of time (generally one month) between standard output and actual output for the inputs used. The purpose of the report is to make the invisible events of small repetitive losses visible through aggregation. In other words, the visibility is attained through the passage of time and the accumulation of small but frequent efficiency losses. The aggregated variance number, if large enough, becomes the signal for management attention.

This method of control has a number of very real limitations in contemporary environments. Table 1 (panel a) provides a listing of the disadvantages of engineering control.

LATE

Often the reporting system will capture and report data on a monthly basis. This means the ability of operating managers to control their processes is severely limited by the reporting frequency of the control data. A monthly report will not give the manager the feedback necessary to change the operating characteristics of the process in the time necessary to prevent losses. Indeed, it is doubtful that even weekly feedback of data is sufficient. Losses are the signal for corrective action, therefore losses will occur before correction is implemented. This is retrospective management.

Figure 1. The Engineered Control Model

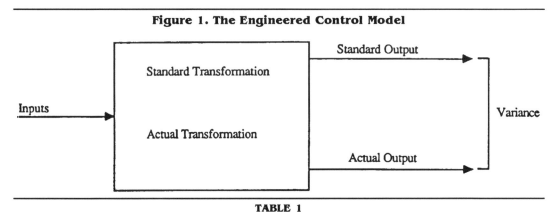

TABLE 1
The Attributes of the Engineering Control and Statistical Process Control

Panel a. Weaknesses of Engineering Control	Panel b. Strengths of Statistical Process Control
1. Not timely	1. Real time data
2. Fails to promote process improvement	2. Promotes continuous process improvement
3. Focuses on ends measures of performance	3. Focuses on causal factors of performance
4. Myopic—production oriented only	4. Multi-factor oriented, productiviy, quality, time
5. High level of data aggregation	5. Time order of data preserved
6. Top-down narrow control	6. Horizontal and process control
7. Performance levels, ignores issues of variation	7. Performance bands and variation highlighted

To illustrate further, suppose the cost system reports an unfavorable $1,000,000 material usage variance for the month. There are two responses to such a report, both of which are symptoms of poor management. First, management may respond that this condition was already known, and that the management reports are merely confirmatory. In this instance, the admission that the condition was known is also an admission that corrective action was not taken, either because the causes were not known or that the control strategy was so ill defined so as to prevent correction. Second, the report may be a surprise to management. In this instance management must wait for the reports before action can be taken. Unfortunately, the variance report is an ends measure of some causal influence. The causal influence is totally obscured. There is no data by which to determine exactly why the material losses occurred. In both scenarios, proactive management is missing. There is no opportunity to prevent the loss from occurring initially.

FAILS TO PROMOTE IMPROVEMENT
As a further indictment, the engineered standard cost system also fails as a motivator of process improvement. As stated at the outset of this paper, the job of management is to continuously improve processes. Yet standard cost systems essentially influence the organization toward a much lesser objective, just meeting standard. Where is the motivation for continuous improvement? Continuous improvement is a philosophy of improving against actual in period t-1. It is not just meeting an engineered standard that was established sometime in the past.

Engineering control is particularly harmful when the standard cost system builds waste into the standard, and thereby conceals the improvement potential. A major wood products firm decomposed their material standards only to discover that there was a 20% material waste built into the standard transformations from raw material to finished board. This amount represented a very significant dollar savings potential. Management was completely surprised by this discovery. No one had ever decomposed the standards, therefore no one really knew the extent of cost improvement potential existing in the enterprise.

Firms that are beginning to work on continuous systems improvement soon discover that the system for updating standards is unable to track the improvement work. The engineering department does not have systems designed to update standards on a frequent basis. The result is accounting systems that are always lagging the engineered reality. The solution is to replace the engineered standard with a rolling actual cost and, thereby, release engineering effort away from administering standards toward supporting systems improvement.[1]

ENDS MEASURE
The variance information that comes from the standard cost system is, by design, an ends measure. The detailed reporting of efficiency information serves only as a control function, not as a corrective tool. A control function means that the data is useful for assessing the past and determining if past events are within engineering expectations. If past events are not within these expectations, namely there are significant unfavorable variances, then the ends measures provide no information as to corrective action. The causal factors are what are important in the corrective phase. Engineering control feedback is as useful as are semester report cards. The report card will communicate the result, but if the result is

[1]An example of the use of rolling actuals is Hewlett Packard as explained in McNair et al. (1988).

undesired it gives few clues as to the underlying system abnormalities that promoted the poor performance. Such reports are only useful for management
oversight of operations, but much less so for the daily operating and improvement
of engineered processes (Kaplan, 1988).

MYOPIC

An organization that controls factory processes through the imposition of
engineered standards has made an assumption about the operating environment
that may be false, or at least limited. That assumption is that the lowest delivered
cost can be obtained by maximizing the production efficiency of all the sub-factors
of production. There are at least two potential problems with this perspective.
First, it is becoming increasingly clear that the performance of systems is multidi-
mensional. Specifically, achieving the most output for the least input can lead to
dangerous conflicts on dimensions of quality, deliverability, and service. The
transformed product in the cost center must meet the quality demands of the
downstream customers (and final customer) in order to be meeting overall firm
objectives. The emphasis of engineered standards on production can subsume the
importance of quality.

An example is a machining operation in a major defense contractor. The
machining operation required both ID (inside diameter) and OD (outside diameter)
grinding to specifications. The grinding operations were under engineering control,
therefore there was an incentive to work quickly in the grinding operations. The ID
grinding was on a ring and the OD on a shaft that fit the ring. Specifications were
developed using statistical tolerancing of the fit pieces. The operators would grind
the OD so it would just meet the upper specification diameter limit and the ID so
that it would just meet the lower specification diameter limit. Both actions minimized the grinding time and provided for excellent efficiency ratings for this
operation. Unfortunately, the fit of the two pieces in a downstream assembly produced a 25% rework rate. This occurred because the diameter distributions of the
two pieces were not centered in the specs, but were opposing each other at the
borders of the specs.

Secondly, the assumption that the summed sub-factor efficiencies will lead to
global efficiency is seriously under question. Consider the experience with JIT
systems. In these systems the engineered standard is not the mechanism that
paces the work. The work is paced by the demand of downstream processes,
which themselves are linked to customers. This is the essence of "pull" vs "push"
production. Push production requires the efficient use of resources by imposing
standards on all factors of production. However, the global costs of inventory control, handling, scheduling, and tracking are ignored under this assumption. The
WIP inventory that is built up by one cost center in the striving for efficiency is lost
by the organization through the increased costs of managing excess in-process
material. Another example is the case of the bottleneck operation. If all workcenters are required to produce at peak efficiency, then the off-bottleneck operations
simply produce inventory and not throughput. This, of course, is the lesson of *The
Goal* (Goldratt and Cox, 1986). Namely, the bottleneck is the pacing operation of
the facility, and that it is unproductive to run the facility with each workcenter at its
own unique pace.

The engineered standard assesses work of people and machines on the basis
of work completed. There is little room for slack operation. Yet slack operation
may be totally appropriate for an off-bottleneck operation or a workcenter in which
there is no downstream demand. This suggests a much different role for line

workers. The slack time is the time for working on process improvement issues, yet the engineered controlled environment can be inflexible to this requirement.

DATA AGGREGATION

Probably the most serious shortcoming of engineered controlled processes is the aggregation of time ordered data into an accumulated variance of performance over time. The loss of time order in the data is critical. The data behavior over time gives important evidence about the *causes* of process behavior. If the process is losing efficiency, it is helpful to know how and when this is happening. Is the loss in efficiency a gradual loss over time, is it immediate, or is it due to a single one time event? These are the kinds of questions that can be answered when the time order is preserved. Almost all significant causal influences such as material lot, fatigue, shift effects, machine wear, ambient temperature, personnel assignment, and the like are related to points in time and can be discovered through an investigation of time ordered data. Engineering control does not report this record and thereby denies the analyst the major tool of process improvement.

TOP-DOWN AND NARROW CONTROL

One of the tasks for the accounting community is to reconcile the changes that will come about from using high performance work systems and employee involvement with the needs of controlling the organization. The engineered cost center worked well in the day of repetitive and process focused manufacturing. The manufacturing organization of today is moving away from such an orientation. Work is being organized in cross-functional teams around the dedicated product family as opposed to remaining specialized in a dedicated manufacturing process. This means that the definition of the responsibility center will expand across a product line and the imposition of control from management will become much less pronounced.

Firms that are experimenting with employee involvement are giving workers much greater ownership of the systems. Ownership implies trust. Engineered standards that are imposed and monitored from above are not consistent with such a philosophy. Ownership requires that the processes perform on multiple dimensions, and that performance measurement expand beyond individual efficiency ratings. Performance will be more and more related to global performance. Much of the detailed information about the process will stay at the process level and will be determined by the process team. There will be less of a need for all information to be centrally accumulated then redistributed. The feedback process will be much closer to those who need the information for process improvement and control. The central data collection and reporting process will become much less focused and broader in scope.

One of the toxic consequences of engineering control is promoting fear. If the employee is disciplined or rewarded on the basis of efficiency ratings, then there is a behavioral incentive to change the data. There are countless examples of this kind of behavior. Indeed, a great deal of engineering and accounting time is spent discovering the causes of data irregularities. The environment where the employee owns the micro system allows the truth in the data to be revealed and, therefore, promotes timely system intervention.

Another well known problem with narrow responsibility reporting is the improper tracing of responsibility to the center. Frequently, the causal factors that lead to reported inefficiencies within a cost center have nothing to do with the cost center's performance. The ends measures in one engineered center can be far removed from the causal factors emanating from another responsibility center, so that there is a great deal of misallocated and poorly captured joint responsibility. For example, machine or labor inefficiencies within a responsibility center can be caused by the condition of equipment (maintenance system), material quality (purchasing system), number of set ups (scheduling system), product workflow (routing system), tool quality (tool replacement system), and management attitude. Indeed, the inefficiencies caused by the individual employee is probably very small relative to the cross system effects. Naturally, this can be very discouraging when the responsibility far exceeds the authority for corrective action. This ability and incentive to pass hidden (non-measured) system failures to other cost centers is probably one of the most dysfunctional consequences of the engineered management approach. A team based systems orientation, allowing for ownership and the power to work on cross-functional systems is the key for eliminating manufacturing waste. The information and control system for such an environment will be much more horizontal, selective, and simple (Cole, 1985).

LEVELS FOCUS

Most performance measures focus unduly on the level of performance. In the case of engineered cost centers the level of performance is measured against the standard level. Such an orientation can cause the firm to ignore issues of variation. Variation is also a contributor to cost and organizational ineffectiveness. Process managers must have an understanding of both the inherent variability in the process as well as the level of process performance. The remaining part of this paper will deal with the importance of variation in understanding and improving processes.

STATISTICAL PROCESS CONTROL

Statistical process control (SPC) is beginning to replace the engineered cost center. The impact of this tool on accountants will be profound. Many of the elaborate controls now existing in organizations will be reduced considerably as SPC begins to replace traditional control devices.[2] The disadvantages of engineering control identified in the previous section are minimized to a great extent using SPC methodology. Table 1 (panel b) provides a listing of the parallel advantages of SPC. These advantages are described and illustrated in the sections to follow.

THE CONCEPT OF STATE OF CONTROL

The original use of statistical process control (SPC) was to improve the output of manufacturing systems. Firms are beginning to recognize that SPC is also a management tool (Deming, 1986; Ishikawa, 1985). Systems improvement work requires "getting into the data." The nature of the underlying variation in the data gives insights to that not available by any other means.

The concept of state of control is the beginning of systems improvement work. Figure 2 illustrates three different control charts on an output for a particular system. The control chart preserves the time order in the data, and therefore provides the "audit trail" for improvement work. The center line is the average value for the process, while the upper and lower control limits represent the three standard deviation limits from the center line based upon the short term variation

---

[2]See Reeve and Philpot (1988) for examples of SPC used to control and improve financial processes.

Figure 2. Three Control Conditions

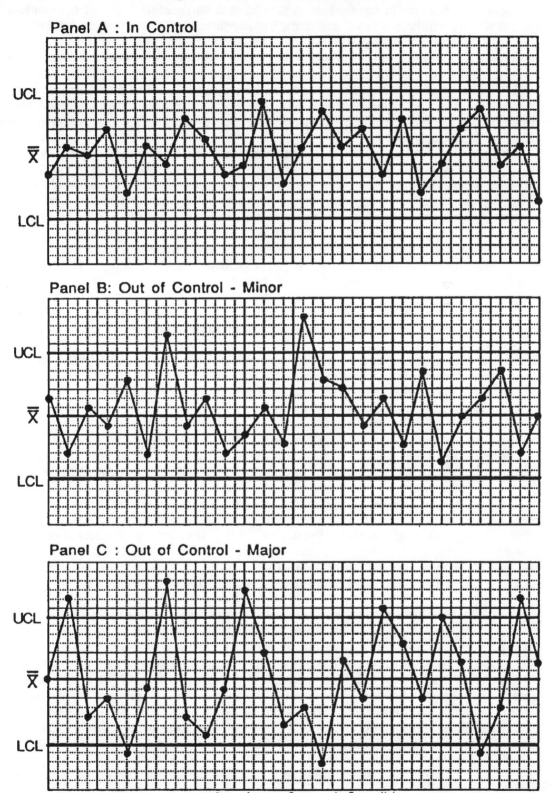

in the process.[5] The top panel is a system that is said to be in control, meaning that the variation inherent in the system is predictable and stable. State of control is a necessary prerequisite before initiating improvement on the common cause system. Without control the system is in a state of chaos and is unpredictable.

The second and third charts are two systems that are out of control. The points outside of the control limits are caused from abnormal events or influences that lead to erratic and unpredictable behavior. These out of control points are not part of the common cause system, but have special identifiable causes. Notice that these points are hidden and unknown in traditional engineering control, because variation is not measured.

The management use of control charts begins with distinguishing between the two types of variation, special cause and common cause. Special cause variation requires a "fire fighting" response. Frequently, system team members should have authority to correct and attempt to eliminate special cause variation. Once the process is brought into a state of control the team can begin the serious work of improving system performance. A controlled process can only be improved by a direct intervention strategy. The intervention strategy is then assessed with control chart data. In the language of SPC a process under control is not necessarily a desired process, it is only a predictable process. Achieving desirability requires management effort.

State of control is important to the accountant for at least two reasons, First, a process that is not in control is also not predictable. Unless the material yield, operating rate, and workloads are predictable then there can be no cost control. This means that attempts to economically model uncontrolled processes can be dangerous. For example, if the machine cycle time is not in a state of control, then attempts to predict the cost of production from such a machine will be hampered by the lack of inherent operating stability.

Second, the performance appraisal system should recognize the distinction between in and out of control. Individual or system performances that are inside the control limits are part of the system (Scherkenbach, 1985). This means that better than average performance in one period does not necessarily signal reward. (Un)exceptional performance is statistically determined by being consistently (below) above average or by being "out of control." This line of reasoning suggests that the accountant should not be thinking in terms of performance levels, but in terms of performance bands. Performance within the band is consistent with the system of selecting, training, and developing people. Performance that is outside the system requires special treatment (discipline, reward). The experience at Ford was that no more than 5-11% of the people were outside of the system.

THE RELATIONSHIP BETWEEN COST AND VARIATION

For purposes of cost control there are two types of variation that are critical. Product variation and process variation. Product variation can be variation within the product, such as dimensional variation, or variation across the product line. The variation across product lines is an issue discussed extensively by a number of authors with respect to identifying complexity cost drivers, so will not be discussed here (Cooper and Kaplan, 1988; Shank and Govindarajan, 1988). Process variation is the variation within systems, and is frequently expressed in terms of units of time.

---

[5]The details of calculating control limits are not discussed here. See Borden (1988) for some examples and discussion on calculation details.

INTRA PRODUCT VARIATION

A generally untapped source of great cost savings is in the area of material usage. This is especially the case with material intensive enterprises such as textile, paper, chemical, and food processors. The accounting treatment of material waste or usage variances is, as stated previously, retroactive. Process measurement is coarse and essentially non-actionable. Material usage variances are not useful for control decisions. Their usefulness lies mainly as a historical report of aggregated events. Manufacturers attempting to use variance reports for operating control will be so far removed from causal factors as to make the work of improvement highly ineffective.

Making the Invisible, Visible in Real Time

What is needed is a measurement system that makes the invisible events of small material losses visible by means other than aggregation. This is the approach of SPC. Take for example the pack and fill line of a consumer goods company. The pack line fills material into a package that is labeled to hold 16 oz. of product. The control chart in Figure 3 provides the weight measurements of packages sampled from the line.

There are two important observations from this data. First, the average amount of weight in each package is 17.5 oz. The reason for this average amount of weight is due expressly to the amount of variation in packing material from the pack heads. The amount of variation forces the average pack well above the lower spec limit. The lower spec limit must be protected. Labeling requires that at least 16 oz. of product be contained in each package. The regulatory cost of underpacking is high, therefore, variation translates directly into higher average material weight. The greater the variation the greater will be the amount of excessive material.

This is a classic scenario for building waste into the standard. The conventional logic asserts that the pack heads have a limited statistical capability in terms of pack weight variation. The standard should incorporate this lack of capability since

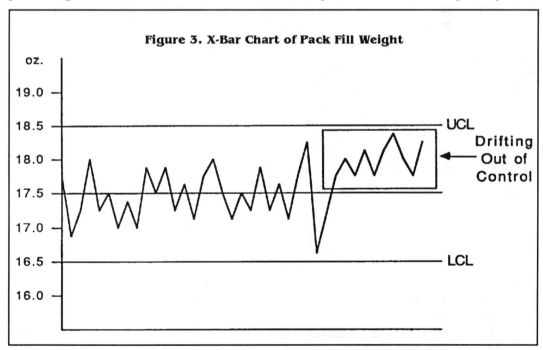

Figure 3. X-Bar Chart of Pack Fill Weight

the packing department manager would receive large unfavorable material loss variances for conditions under which there is no control. The flaw in this line of reasoning is that the standard has built in a very large amount of material waste that should be made visible. The visibility of the waste can lead directly to operating decisions to eliminate the waste. Decisions on changing the repair cycle on the packing machines to decrease variation, improving pack head performance through capital investment, or improving upstream material processes are examples of operating decisions that can be made from this data.

Indeed, this is exactly what this company did. The discovery of the pack head variation lead the process improvement team to identify through the time order of the data the root cause, which was the consistency of the material being packed. As a result changes were implemented in material processing so that material consistency would be more uniform, thereby allowing the pack to have much lower variation. The lower variation translated directly into a shift down in the average pack and a direct cost savings. Such an improvement was not even considered from the traditional management report. Moreover, the ability to move up the system required a different management organization. The employees were part of a product team that had responsibility for the complete product line. This allowed much stronger ownership and horizontal control of the system.

The second important observation from the SPC chart in Figure 3 is the trend that is developing from the data. This trend is seven consecutive points above the center line. Seven consecutive points is an accepted rule for determining an out of control condition. This means that there is statistical evidence that the mean of the process has shifted for some reason related to the time period sampled. The mean is now above 17.5 oz. The difference between the new process average and the old average is the basis for determining a *prospective variance.* The prospective variance is the amount of additional material losses above a 17.5 oz. standard that will occur over an assumed time period if the process is allowed to continue under its present state.

Management now has the ability to make an economic decision that will affect future cost. The operating responsibility is to accumulate sufficient data to determine why the pack weights begin drifting upwards. This analysis may reveal that the cause is related to accumulation of material in the pack heads. Operating personnel can begin using the statistical and financial support data to develop optimal pack head clean out policies that will simultaneously minimize material losses and downtime. Again, this type of analysis is generally not possible from traditional material loss reports.

The above example is one of many that could be related. As further examples, the variation within products helps managers understand the material usage of coatings on paper, the amount of trim on wood products, the dimensions of discrete parts (which impacts assembly capability and rework), the amount of carpet fiber add-on in carpets, the amount of lubricant used to make poly extruded yarns, and the yield of chemical processes. In all of these the use of statistical methods reveals process knowledge that is simply missing from traditional variance reports. As firms begin to embrace statistical techniques the use of traditional material usage variance reports as a control device will become increasingly anachronistic.

PROCESS VARIATION

Process variation exists in the strategic and operating systems of an enterprise. The impact of variation on these systems is to increase cost, restrict throughput, increase capacity, and increase coordination costs. Both the macro and micro systems can exhibit variation.

Strategic Systems Variation

Strategic variation is the variation in the macro systems of the enterprise. Such systems have a pervasive impact on the organization, and as such, provide an excellent opportunity for improvement. However, these systems are also difficult to change since their ownership is highly distributed.

An example of strategic variation is the fairly typical decision to run cents-off promotional campaigns in consumer product companies. The impact of this decision is illustrated in Figure 4. The effect of the promotion campaign is to effectively transform a predictable and level demand into a highly variable production schedule by accelerating demand into the promotion time period. Essentially, demand variation not existing inherently is manufactured through marketing decisions. The costs of these decisions are presently hidden in the overhead structure of many firms, but are real nonetheless. The costs are incurred in the form of higher overtime and capacity costs to meet the surges in production; greater costs for filling and managing the material and distribution pipeline; and additional direct and indirect packaging costs for special cents-off labeling.

What is missing is balance in the decision. The role of accounting is to properly trace these variation costs to products and managers that cause them. For the example above, a cost per promotion could be assigned to the brand being promoted (scaled by the size of the promotion). As a result, the relevant cost of product can be used to build profit responsibility at the brand manager level. In this way behaviors can be modified so that marketing decisions will capture both demand and cost implications simultaneously.

Operating Systems Variation—Processing Times

The operating systems of the enterprise are the micro production and administrative systems that work to perform the everyday tasks of the enterprise. Variation inherent in these systems impact the effectiveness and efficiency of system performance. As mentioned previously, traditional control and productivity measures generally ignore the impact of variation in evaluating system performance.

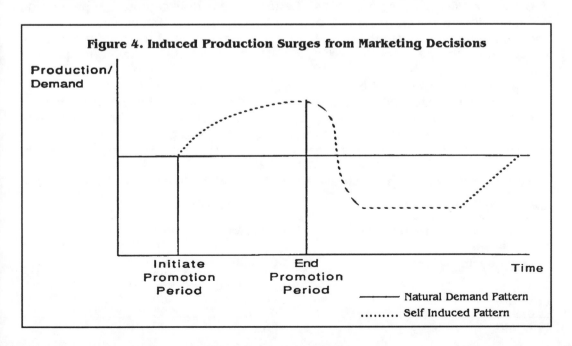

Figure 4. Induced Production Surges from Marketing Decisions

Consider a production line with three consecutive workcenters, each requiring five minutes of processing time to process a single unit. The standard is set at 12 units per hour in each machine center. Machine efficiencies are calculated and reported. The efficiencies for each workcenter will indicate 100% efficiency if the workcenters are able to maintain the *average* processing time of five minutes per unit. Assume that the machines are able to perform to this average, but that the processing times are variable. Variable processing times have considerable consequences to the global effectiveness of the line.

The assumptions above were modeled in an XCELL+ (Conway et. al., 1987) production simulation in order to determine the impact of processing time variation. Conceptually, variation in processing times will cause interference in the throughput of the line. Essentially, processing time variation causes a balanced line on average to become unbalanced during intervals of production. As a result, adjacent workcenters will experience systematic starving and blocking and will be unable to attain full throughput potential.

In the simulation results in Table 2 the simulated factory was run for 1,500 minutes. When the processing times across the three workcenters is held constant at five minutes the throughput potential of 300 units (1,500/5) is achieved. The results indicate degrading in throughput as processing time variation is introduced. Greater variation causes throughput to decline below the theoretical capability. If the processing times vary uniformly between .1 and 9.9 minutes the throughput will decline to 255 units in 1,500 running minutes.

The classic approach to preserving throughput under such a scenario is to buffer adjacent processes with work in process inventory capacity. The simulations were rerun with a buffer capacity of three units between workcenters one and two, and two and three. The buffering of processing interference markedly improved the factory's performance as indicated in the table.

The WIP buffering allows the workcenters to approach full efficiency. However, this is not an optimal solution. The imposition of WIP buffers adds cost in terms of space, capital, movement, handling, and administration to the production process. The traditional accounting measures of effectiveness only capture the production efficiency inside the workcenter, but fail to capture the impact of workcenter variation on the total system costs. A statistical evaluation of the workcenter cycle times will reveal the inherent variation and will provide the baseline for process improvement.

This approach was used by a major paper manufacturer to reduce the amount of in-process inventory stored between the paper making and paper converting operations. This manufacturer discovered that the run rate in paper making was highly variable and dependent upon the quality of the pulp. The converting

TABLE 2

The Relationship Between Cycle Time Variation, Buffer Capacity, and Throughput

Cycle Times for Three Adjacent Workcenters (Uniform distribution between values in parentheses)	Throughput	
	No Buffer	Buffer Capacity 3 Units
Constant 5 minutes	300 units	300 units
(4.5, 5.5)	294	300
(4.0, 6.0)	289	300
(2.5, 7.5)	275	300
(1.0, 9.0)	261	295
(0.1, 9.9)	255	290

operation was highly variable and dependent upon the reliability of the converting lines. The combination of these highly variable processes required the expansion of the parent roll inventory between these two processes to prevent throughput reductions. The manufacturer began working on the cause factors driving variation and was able to slash the parent roll inventory by 80%.

Maintenance Variation

The impact of variation on operating systems has many examples. Another includes the impact of maintenance downtime across the production facility. The net effect of downtime uncertainty is to introduce variation into the process. Indeed, the major source of processing time variation illustrated above are the unexpected downtimes due to maintenance. These unexpected interruptions require buffer inventory to protect throughput. Other sources of variation include randomness in the time between failures and randomness in repair times.

To illustrate the impact of maintenance variation an XCELL+ simulation was developed. The modeled facility has three sequential workcenters (W1, W2, and W3). Constant processing times of four, three, and four minutes are programmed into the three workcenters, respectively. Essentially W1 and W3 are constraining resources. W2 is unreliable and requires maintenance. As a result, buffer capacity is required between W1 and W2 (B1), and W2 and W3 (B2) in order to protect throughput. Table 3 illustrates the impact of key maintenance variables on the operating performance of the facility.

To begin, a fixed maintenance schedule is assumed wherein maintenance occurs every 150 minutes and repair times are constant at 50 minutes per repair. Under this scenario, full throughput is achieved by building buffer capacity in B1 and B2 of 13 units (50/4). As Table 3 shows the throughput for 5,000 running minutes is 1,250 units (5,000/4), which is the maximum potential of this facility. If the assumption of fixed repair times are relaxed so that the repair times follow an exponential distribution with a mean of 50 minutes there will be a loss of throughput. This is so because repair times longer than 50 minutes will not be buffered by 3 units capacity. More buffer capacity is required to protect the longer repair times. Rerunning the simulation with 20 units of buffer capacity in B1 and B2 recaptures

TABLE 3
The Impact of Maintenance Variation on Operating Variables

Maintenance Characteristics

Time Between Failures	Repair Times	Throughput (5,000 minutes)	W1 Utilization	B1 B2 Capacity
Constant 150 minutes	Constant 50 minutes	1,250 units	100%	13 units
Constant 150 minutes	Exponential 50 minutes	1,199	96	13
Constant 150 minutes	Exponential 50 minutes	1,236	99.2	20
Exponential 150 minutes	Constant 50 minutes	1,140	91.2	13
Exponential 150 minutes	Constant 50 minutes	1,155	93.4	26
Exponential 150 minutes	Exponential 50 minutes	1,078	86.2	13

much of the lost throughput, but at the cost of longer lead times and inventory related costs. This is the economic tradeoff the accountant must help answer. Namely, is it more economically justified to work on the source of variation or to increase inventory buffer capacity? The easy solution is to allow inventories to grow in order to buffer the variation, however, there are costs to this solution that must be recognized.

The simulation was next run with constant repair times, but with random time between repairs. The repairs no longer follow a fixed schedule, but follow an exponential distribution with a mean of 150 minutes. The cost of random repairs through lost throughput and equipment inefficiency is severe. The randomness in W2 reliability forces the starving of processes due to the interference between closely adjacent repairs. Close repairs give insufficient time for B1 to completely empty. As a result, only partial buffer protection is achieved when repairs happen close to each other. In this facility the repairs must be at least 150 minutes apart in order to allow enough time for the buffers to clear (12.5/ (.33-.25)). Adjacent repairs closer than 150 minutes apart will result in lost throughput. An interesting observation is that no amount of additional buffer capacity will relieve this situation. Running the simulation with twice the buffer capacity will not improve matters. The only way to avoid lost throughput is by spreading the incident of repairs and by moving closer to a fixed schedule.

Additional costs imposed by random breakdowns include uneconomic scheduling of scarce maintenance support, an inability to use downtime for setting up, expensive schedule interruptions, and excessive scrap due to unexpected shutdown/startup. The cost to the enterprise for unexpected downtimes goes beyond lost equipment efficiency and throughput, but again includes spillover effects into the system.

Randomness in both repair times and times between repairs is, as expected, the worse scenario. Following the discussion above, buffering the process will be only partially successful. Again, the accountant should be in a position to help identify the cost and benefits of reducing variation. Such work will require capturing cost and throughput effects. Tools such as factory simulations are one way to begin to understand relevant relationships.

Other examples of operating system variation that translate into reduced system effectiveness are variations in batch sizes, repair times, time between repairs, raw material and finished goods shipments, schedule, and speed. Most traditional measures of effectiveness and efficiency fail to measure and isolate the impact of variation on processes, and so therefore fail to provide the means for proactive improvement work. Moreover, accountants are just beginning to recognize the cost of variation on systems. Isolating the cost of variation is the beginning of setting priorities in improvement effort.

CONCLUSIONS

In this paper I have provided arguments favoring the use of statistical process control as a methodology of controlling and improving processes. The extensive use of variance analysis was argued to be inadequate to the needs of the world class firm. The ability for SPC to measure both the level and variation inherent in the system was identified as a principal advantage to the statistical approach. The cost of variation was illustrated for the case of processing time variation and variation in maintenance. In both instances traditional accounting measures would fail to attribute the source of cost to the variation.

REFERENCES

Borden, J. P., "Using Statistical Process Control to Become a Quality Manufacturer," *CIM Review* (Summer 1988), pp. 6-18.

Cole, R. E., "Target Information for Competitive Performance," *Harvard Business Review* (May-June 1985), pp. 100-109.

Conway, R., Maxwell, W. L., McClain, J. O., and Worona, S. L., *User's Guide to XCELL+ Factory Modeling System* (The Scientific Press, 1987).

Cooper, R., and Kaplan, R. S., "Measure Costs Right: Make the Right Decisions," *Harvard Business Review* (Sept-Oct 1988), pp. 96-105.

Deming, W. E., *Out of the Crisis* (MIT-CAES, 1986).

Goldratt, E. M., and Cox, J., *The Goal* (North River Press, 1986).

Ishikawa, K., *What is Total Quality Control?* (Prentice-Hall, 1985).

Kaplan, R. S. "One Cost System Isn't Enough," *Harvard Business Review* (Jan-Feb 1988), pp. 61-66.

McNair, C. J., Mosconi, W., and Norris, T., *Meeting the Technology Challenge: Cost Accounting in a JIT Environment* (National Association of Accountants, 1988).

Reeve, J. M., and Philpot, J. W., "Applications of Statistical Process Control for Financial Management," *Journal of Cost Management* (Fall 1988), pp. 33-40.

Scherkenbach, W. W., "Performance Appraisal and Quality: Ford's New Philosophy," *Quality Progress* (April 1985), pp. 40-46.

Shank, J. K., and Govindarajan, V., "The Perils of Cost Allocation Based on Production Volumes," *Accounting Horizons* (December 1988), pp. 71-79.

INTERNALLY FOCUSED ACTIVITY-BASED COST SYSTEMS

Robin Cooper
Associate Professor
Harvard University

Peter B. B. Turney
Tektronix Professor
Portland State University

ABSTRACT

This paper examines three firms which used activity-based cost systems to affect internal decision making as well as to determine product costs.

Activity-based cost systems are most widely known to be more complex and provide more accurate costs than traditional cost systems. They may also be used, however, to promote behavioral changes by using drivers that are linked both to cost and to factors that lead to improvement of manufacturing capability. This internal focus of the cost system promotes product design and process changes that will support a company's strategic plans.

Internally focused activity-based costing systems tend to be simpler and less accurate than their externally focused counterparts. Selection of an appropriate system should be determined by a cost-benefit analysis that considers competitive pressures, product life cycle, and need for accurate costs.

Several recent papers have emphasized the more accurate costs reported by activity-based cost systems (for example, see Cooper and Kaplan, 1988a; Cooper and Kaplan, 1988b).[1] Activity-based systems emerged in firms facing increased competition, where management decided that the fastest way to become more profitable was to gain a better understanding of what it cost to make their products. The more accurate product costs enabled management to take action on pricing and product mix to increase overall profitability. Thus, the systems were designed to lead to modifications in the *external* marketing strategy of the firm.

Recently, another way to take advantage of activity-based systems has emerged. In some firms, management has decided that the fastest way to improve profitability is to reduce product costs through improved design and more efficient processes. These firms are characterized by products with relatively short life cycles; thus, the firms

---

Reprinted with Permission from 'Measures for Manufacturing Excellence,' edited by Robert S. Kaplan, The Harvard Business School Press, 1990.

[1]Activity-based cost systems are distinguished from their conventional counterparts by the cost drivers, or allocation bases they use to relate the consumption of inputs to products. Conventional systems rely solely on unit level cost drivers, such as direct labor hours or dollars, machine hours, and material dollars. A unit level cost driver assumes that inputs are consumed in direct proportion to the number of units produced. Activity-based systems, while retaining unit level drivers, also use non-unit level ones. Non-unit level drivers assume that costs are not consumed in direct proportion to the number of units produced. Examples of these drivers include number of part numbers, number of vendors, and number of setups (see Cooper, 1989).

quickly reap benefits if new products can be manufactured efficiently. They use activity-based systems to send specific messages to product designers and process engineers about how to improve the manufacturing capability of the firm. Their systems are designed to lead to modifications in the *internal* strategy of the firm.

We identified and visited three companies all in the electronics industry, that use internally focused activity-based systems. The three firms produce a diverse mix of products, have high fixed overhead, and have recently implemented continuous improvement programs in response to increased competitive pressures.

We first describe the activity-based cost systems developed in each company. We then analyze our findings; particularly, the conditions that favor internally as opposed to externally focused systems.[2] These conditions include: 1) established competitive strategies and 2) products with short life cycles. The structural differences between the two types of systems are explored, especially the number and nature of cost drivers.

The analysis of the difference between the systems relies upon relatively few examples. We know of fewer than ten documented activity-based cost systems. Our analysis is a preliminary one.

THE EVIDENCE

The three companies illustrate how activity-based cost systems can be designed to motivate improvements in the manufacturing capability of the firm. The primary objective of the system at Portable Instrument Division of Tektronix was to reduce the number of unique components in their products. Hewlett Packard's Roseville Network Division objective was to improve the design of its products via better cost/performance tradeoffs. At Zytec, the objective was to reduce the elapsed time from the ordering of components to the shipment of the finished product.

Tektronix: Portable Instruments Division[3]

Portable Instruments Division (PID) produces portable electronic oscilloscopes. It recently implemented a continuous improvement program to help make it competitive with Japanese products. The program consisted of just-in-time production, total quality control, and shop-level personnel involvement. The changes in the production process resulting from this improvement program, coupled to a significant decrease in the direct labor content of its products over the last few years, helped make PID's direct labor-based cost accounting system obsolete.

Therefore, management initiated a special study to identify the factors responsible for overhead. The study determined that half of all overhead costs related in some way to the number of different part numbers handled. To trace this overhead to products, management selected the cost driver, "number of part numbers." This driver divided overhead costs by the number of different part numbers used in the facility. Dividing the cost per unique part number by volume of each part number used created a charge that was assigned to each product, depending on the overall usage of that part.

Management believed that using the number of part numbers as a cost driver would increase the product designers' awareness of the high costs associated with using low volume parts. This awareness was important because the designers had

---

[2]Our analysis is based in part upon findings from earlier studies on externally focused systems. See for example, the following Harvard Business School case studies: R. Cooper, "Schrader Bellows," 9-186-272, 1988, R. S. Kaplan, "John Deere Component Works," 9-187-107, 108, 1987, and "Kanthal," 9-189-129, 1989.

[3]The material for this section is taken from Robin Cooper and Peter B. B. Turney, "Tektronix: Portable Instruments Division," 9-188-142/3/4. Boston: Harvard Business School, 1988.

used many special components to achieve product functionality; thus the total number of different part numbers used by the division was very high. Management believed it could reduce overhead costs by achieving a reduction of both the number of part numbers in future products and the range of part number-related activities such as maintaining bills of material and expediting parts.

The management considered the burdening[4] approach a success by management and the product designers acknowledged the beneficial effect of the system on their decisions. For example, according to the engineering manager:

> The way we allocate manufacturing overhead costs affects the way we design products. We design products differently under material burdening from the way we did under the old system, where all overhead was charged to labor. With material burdening, we now recognize the acquisition cost plus the carrying costs of components. This knowledge affects us during the design process, automatically guiding the process to the minimum true cost of doing business, not just to the lowest parts cost.

> Today, we take a hard look at what parts we really need and deliberately minimize the proliferation of equivalent parts. Three major factors force us to limit part numbers: automatic assembly equipment can only handle a limited number of parts; each additional type of part complicates the task of establishing and maintaining high quality sources of supply; and each part type complicates the logistics and cost of operating the production process.

Hewlett Packard: Roseville Network Division[5]

The Roseville Network Division (RND) of Hewlett Packard produces networking devices that allow computers to communicate with other computers and peripheral devices. The rapid introduction of technology in the last few years has dramatically reduced the life expectancies of RND's products, while increasing the number of different products it has to sell to remain competitive.

RND's strategy is to keep the products up-to-date and to design them in a cost effective manner. The shorter lives and increasing number of its products require a continuous flow of new products from design to production. Careful attention to the design of the products to reduce their overall cost of manufacturing is also needed.

The goal of the new cost system is to encourage the product designers to choose the least expensive alternative between designs with equivalent functionality. The cost system provides cost information on each production process so that the costs of different designs can be compared. A manager deeply involved in the implementation of the new cost system stated:

> The purpose of cost driver accounting[6] was not to prevent the engineers from introducing new technology. Rather it was to get the engineers to think about cost, and not to go for elegance every time. Cost driver accounting put product costs on the backs of the engineers. It encouraged them to design for manufacturability.

The initial system design was quite simple: it had only two cost drivers: direct-labor hours and number of insertions. However, over several years a team from accounting and engineering sequentially added new cost drivers to the system as its understand-

---

[4]PID uses the term "material burdening" for the assignment of part-related overhead costs.

[5]The material for this section is taken from Robin Cooper and Peter B. B . Turney, "Hewlett Packard: Roseville Network Division," 9-189-117. Boston: Harvard Business School, 1989.

[6]RND uses the term "cost driver accounting" to refer to its activity-based cost system.

ing of the economics of design improved. For example, early in the development of the system an engineer believed that the reported cost of a product was too high. A special study confirmed this belief and showed that axial insertions were about one third the cost of dual in line processor (DIP) insertions. The cost system was modified to differentiate between the two types of insertion.

During a four year period the cost system went from two drivers to nine. The first six of these drivers are surrogates for direct labor or machine hours. These unit-based drivers, however, are much more meaningful to product designers. Designers find it easier to think about the number and type of insertions their designs require, rather than how long they take to manufacture. These drivers were:

1. Number of axial insertions
2. Number of radial insertions
3. Number of dip insertions
4. Number of manual insertions
5. Number of test hours
6. Number of solder joints
7. Number of boards
8. Number of parts
9. Number of slots

The designers did not, however, use the cost data blindly. As one designer commented:

I do not design to minimize the reported cost of the product as this will change across time with the introduction of new cost drivers. Instead, I design what I think is the lowest cost product for a given level of reliability and functionality. It is important not to design to minimize cost because the model does not capture all of the relevant costs.

The cost system affected the decisions made by the product designers by allowing them to develop a series of design heuristics that helped guide their design process. For example:

1. Manual insertion was three times as expensive as automatic insertion.
2. Connectors that could not go through the wave solder machine and were manually inserted, or needed special pre-solder treatment, added approximately $2-3 to the overhead cost of the board.
3. Ease of availability was similarly critical. Selecting a component that had excess production capacity in the industry and could be supplied by multiple vendors reduced the risk of component shortages. The rule of thumb used to adjust for availability was that a low availability component had an additional cost of ten times its material cost.

Once the designers had developed these heuristics, they did not keep going back to the cost system for cost information. However, each time a new cost driver was implemented the designers studied the effect on the economics of production and adjusted their design rules accordingly.

The new cost system thus played an important role in the design of new products. Even though it was based upon historical data, management believed it was successful in changing product design to improve the performance of the firm. The engineering manager at RND described how the design process changed:

We created a lot of tighter relationships between accounting, research and development, manufacturing and marketing. We were all learning about the business. We have broken the back of the cost system design problem and are now refining it and our intuitions about the economics of product design. Overall, the whole experience forced us to understand our design process.

Zytec Corporation[7]

Zytec Corporation is an independent producer of custom power supplies for computer peripherals and other electronic products. It was formed in a leveraged buyout in 1984. Zytec management embarked upon a continuous improvement program in late 1984. It has implemented just-in-time production techniques, total quality control commitment and people involvement programs. The continuous improvement program was extremely successful; for example, over a three-year period the reported cost of one product fell from \$530 to \$325 and the cycle time was significantly reduced.

One of the major elements of this continuous improvement program was a Management by Planning system implemented in late 1988. This system identified six objectives that management believed to be critical to the success of the firm during 1989. All departments were expected to identify improvement targets for these six objectives. These objectives were:

1. Improve Total Quality Commitment
2. Reduce Total Cycle Time
3. Improve Service to Customers
4. Improve Profitability and Financial Stability
5. Improve Housekeeping and Safety
6. Increase Employee Involvement

The changes in the production process brought about by the continuous improvement program coupled with a general decrease in the labor content of the product rendered the existing cost system obsolete. The team assigned to redesign the system was told to design one that reinforced the just-in-time philosophy. Zytec's controller explained why:

> We wanted to pick a set of drivers that was meaningful to the people on the floor. We wanted these drivers to capture the essence of our drive for continuous improvement. In particular, we were convinced that the cost system could become a potent tool for behavior modification. The only real limitation that we placed upon ourselves was that the cost system could not require any special measurements. We simply do not have the administrative resources.

The design team identified four potential cost drivers:

Yield: Yield, a test of the quality of production that measured the average number of pieces that did not have to be reworked. Yield was considered a cost driver because of the high cost and shop floor disturbance associated with rework.

Cycle Time: A measure of the average time taken by a product to go from raw material to finished goods. Cycle time reductions could be achieved only by improving the yield to reduce rework.

Supplier Lead Time: A measure of the average time taken by a supplier to deliver an order to Zytec. Supplier lead time was considered a cost driver because long supplier lead time was a significant roadblock to the flexibility Zytec sought to provide its customers.

[7]The material for this section is taken from Robin Cooper and Peter B. B. Turney, "Zytec, Inc. (B)," 9-190-005. Boston: Harvard Business School, 1989.

Linearity: A measure of the average absolute deviation of actual production to the daily production plan. For example, if the planned output is 10 units and the actual output is either 12 or 8 units then the daily linearity is said to be 80%.[8] Management believed that higher average linearity lowered overall costs.

After a series of discussions with management, the team simplified the cost system design and retained only two of the four drivers, cycle time and supplier lead time. All manufacturing overhead, previously allocated using direct labor hours, was allocated according to cycle time and all material overhead, previously allocated using material dollars, was allocated proportionately to supplier lead time.

The team did not use linearity and yields as cost drivers because it had little confidence that the formulae it had developed for them would correlate with cost. The controller commented:

> We wanted to build a system that fit manufacturing's need. They (manufacturing people) felt that cycle time and supplier lead time were most important, so we constructed a system based on these factors.

Despite limiting the design of the system to the two cost drivers that management felt were most important, the team found reaction to the system predominately negative. One designer commented:

> The initial reaction to the cycle time portion of the new system was dominated by confusion. Everybody wanted to compare the old and new numbers and have us explain the differences. Unfortunately, we could not satisfy them about the causes of the differences. This problem was exacerbated because some monthly reports used direct labor figures based upon industrial engineering standards, while others used the allocated figures which naturally were different.

> Reaction to the supplier lead time system was even more negative. In particular, the purchasing manager argued that supplier lead time was not under his control. He could not see any way to change his behavior to reduce supplier lead time. He felt unable to react to the new system.

Given this negative reaction, the new cost system was modified. Material burdening was treated as a period cost and only manufacturing burden was allocated to the products using cycle time.

ANALYSIS

Activity-based systems, both internally and externally focused, typically appear in firms that are experiencing intense competitive pressure, and where the benefits from the more flexible but more costly and complex activity-based system exceed the costs of designing, implementing, and operating such a system (Cooper, 1988b and 1988c). The three companies studied were all experiencing intense competitive pressure; PID from the Japanese, RND from changes in the product life cycle, and Zytec from a number of competitors. Similarly, they demonstrate the conditions that cause the benefits of an activity-based system to exceed the costs: a diverse product mix and high fixed overhead (Cooper and Kaplan, 1988).

Firms that have introduced internally focused systems differ from those that have introduced externally focused systems. A key difference is the extent to which the competitive strategy of the firm has been clearly established. A second difference is

---

[8]The absolute value of (10-12) and (10-8) is 2. The linearity is therefore 80% ((10-2)/10 stated as a percentage).

the length of the time from introduction to withdrawal of the product. The length of this life cycle affects the orientation of internally focused systems.

The Role of an Established Strategy

Internally focused systems are used to make internal improvements in order to better implement a chosen strategy. They are used to facilitate improvements in product or process design. This was the case in all three of the internally focused systems. At PID, for example, the system was purposely designed to support the firm's strategy. As one of the cost system designers commented:

> We recognized that we had already chosen our strategic direction. What we needed was a cost system that would drive behavior in that direction.

At RND, the strategy was to keep the product lines as up-to-date as possible. The short product life cycles reflected this approach. Management implemented a cost system that motivated the design of products that could be manufactured more efficiently. At Zytec, the strategy was to improve manufacturing capability, with particular emphasis on improving cost, quality and flexibility. The focal point for implementing this strategy was the reduction of total customer lead time. This was the primary motivation behind the design of the new cost system.

In contrast, most firms that implement externally focused systems use the new product costs to help establish a new strategy. The benefits of the cost system to these firms comes from the improved profitability resulting from changing the product mix, product prices, and other factors relating to the firm's approach to its external customers.

The need for an established strategy before introducing an internally focused system is understandable. First, the cost system can be designed to support and enhance the chosen strategy; it can be focused on motivating specific improvements in the firm's manufacturing capability. It can be focused on motivating specific improvements in the firm's manufacturing capability. Second, an established strategy creates a receptive environment for the new system. A system that says the message "reducing the number of unique components, will make it easier to introduce new products" will be better received in a company whose strategy is to continually introduce new products than in a company that has yet to decide if new products are beneficial.

Not all companies with a well-established strategy have internally focused systems. At least one firm, the Electric Motor Works at Siemens has implemented an externally focused one to sustain its new strategy.[9] They chose this route because the new strategy rendered the existing system obsolete. Without a new system they would have been unable to tell which orders were profitable and their new strategy would have failed.

The Role of Products with a Short Life Cycle

All three firms implementing internally focused cost systems had products with a short life cycle.[10] The product life cycle at PID was about 18 months, at RND about 24 months, and at Zytec a somewhat longer 36 months. In contrast, the product life cycles at firms that installed externally focused cost systems are typically much longer. The role of product life cycle in determining the type of cost system to install can be

---

[9]R. Cooper and K. Wruck, Siemens Electric Motor Works (A): Process Oriented Costing, 9-189-089. Boston: Harvard Business School, 1988.

[10]For more information on the linking of manufacturing and product life cycles, see Hayes and Wheelright, 1979.

understood by considering how the length of the life cycle affects the speed at which different benefits can be achieved. If a firm's existing product mix is insufficiently profitable, and the products sold have long life cycles, the fastest way to improve performance is often to change the mix of products sold or prices charged. To make these changes requires understanding relative product profitability.

In contrast, if the product life cycle is relatively short, the fastest way to improve performance is often to design new products that can be manufactured efficiently. Hence designers need to be motivated to design products for manufacturability.

At PID and RND, the primary focus of the cost system was to improve the design of the product. The firms had already made their production processes highly efficient. Management believed that larger cost reductions could be achieved by better designs of the next generation of product than by improving the efficiency of the existing products or production process. The manufacturing manager at PID described this perspective:

> Part number burdening is of little immediate use to the manufacturing manager. Its real value is to drive the next generation of products to have fewer parts. Manufacturing can not do much, in the short run, about the number of parts. The reduction in the number of parts can only be significantly reduced through product redesign.

The focus of the system at Zytec was slightly different. The longer product life cycle increased the benefits from improving the production process for current products. Consequently, the system focused on improving both the current production process and the design of new products.

Simplicity

Internally focused systems tend to be simpler than externally focused ones. Simplicity helps to ensure that the system is understood by management and product designers, and is therefore more likely to produce the performance improvements desired. For example, at PID, management purposely kept the system simple so that the users understood the "message":

> Our objective with the cost system was to change the behavior of the division management. We did not go to a more complex system say an eight driver system immediately, because we wanted to take incremental steps and change behavior permanently. If we had tried to introduce a massive change, we might have created such confusion that the benefit would have been lost. The beautiful thing about taking this incremental step was that the behavior change we wanted was almost instantaneous. On day one we thought the old way, and on day two we were thinking the new way.

However, this approach will only be effective if management can react to the message. At Zytec, for example, the driver, "supplier lead time" was poorly accepted because the purchasing manager did not control supplier lead time and therefore could do nothing about it.

The danger of simplicity is that the reported product costs may be somewhat inaccurate and lead to poor decisions. The message is strong. If the reported cost per unit of a cost driver is higher than it should be, this approach risks improvements being made that are not cost justified. The firms' original direct labor based allocation systems suffered from this problem. As one PID manager commented:

> Engineering could justify a $10,000 project to remove five minutes (of direct labor) from an instrument because that five minutes was leveraged by the labor-

burdening system. In reality, they were adding to the size of burden and doing little to reduce costs.

The designers at all three companies were aware of the risks of reporting inaccurate product costs. One designer at PID commented:

The product costs reported by our system are accurate enough to guide behavior in the direction we want to go. They are not 100% correct nor are they as accurate as we want them. I did not attempt to calculate the economic or financial impact of designers' individual decisions. This was not necessary because collectively their decisions would take us where we wanted to go—common parts, standard parts throughout the product line, and modifications to existing products using existing components. I was more interested in a move towards standardization than being able to say that replacing two unique parts with one common part would save exactly $4.93.

The designers at PID and RND had strategies for dealing with the lack of accuracy of their systems. PID had identified two possible approaches. One was to change the drivers once the message had been accepted. As a manager at PID commented:

The number of parts will cease to be a useful behavioral cost driver when the design engineers see the value of common parts. When they come to believe that it hurts our competitive position to proliferate parts, they will naturally design the products with common parts, and we won't have to continually remind them to do so.

The cost system will be used to cause behavioral change in a series of steps. I can see that in a few years the system will use different cost drivers, for example, cycle time burdening.

An alternative approach was to add more drivers to the system sequentially. This approach was used by RND. In response to engineers' expressed concerns, the system was continuously modified to make it more accurate. An engineer commented on the process of continuously changing the cost system:

I would have preferred one transition, but I do not believe it could have been done that way. Accounting simply did not understand enough about the production and design process. If we had attempted one transition we would have risked freezing the firm on the first system we designed and we would not have been able to change it to reflect new insights we gained from it.

However, this additional accuracy was achieved at the expense of the simplicity of the system. Personnel at RND were concerned that its system was becoming too complex. According to one engineer on the design team:

We could improve it (the cost system) to capture more costs but we would risk it becoming too complex to understand and hence use. This is especially true for new college hires. They have never been taught to design with cost in mind.

The designers at Zytec were less successful in resolving the accuracy issue. Their system was so focused on motivation that management and engineers did not believe in the accuracy of the costs reported. For example, supplier lead time is not a good driver for product costing because it does not capture enough of the complexity of material acquisition costs. For example, one vendor may have a long lead time but deliver high quality products while another might deliver low quality products with a short lead time. The failure occurred when management tried to interpret the new cost numbers in light of the old ones.

When the driver that sends the strongest message is not the one that most accurately reports product costs, designers will have to choose between the strength of the message or the accuracy of the reported product costs. Choosing a driver that does not send the desired message very clearly risks not achieving the desired improvements. Choosing a driver which does not report very accurate product costs risks management making poor decisions based on the costs reported. We know very little about how to decide which type of driver to choose. More research is required before we understand the nature of this trade-off.

CONCLUSIONS

This chapter introduced the concept of internally focused activity-based cost systems. These are activity-based systems designed to influence internal decisions such as product or process design. They can be contrasted with externally focused systems designed to report accurate product costs and allow the firm to modify its strategy.

The firms implementing internally focused systems made products with short product life cycles and had well-established strategies. Those implementing externally focused systems made products with longer life cycles and usually were searching for a new strategy. Between the two types of systems, two structural differences were identified: the number and the type of cost drivers selected. On average, internally focused systems appear simpler than externally focused ones, and use cost drivers that send a clear message in contrast to systems that report accurate product costs.

An important question that emerges from this study is whether cost systems should be designed to influence internal decisions. The authors believe the answer to this question is, "yes." First, at both RND and PID, managers strongly believe that the cost system is a potent method to improve the firm's manufacturing capability. There is also strong evidence at RND that designer behavior was modified by the cost system. Second, ignoring the motivational effects of cost systems can lead to decisions that negatively affect performance. There is evidence at PID and Zytec that the driver "direct labor hours" resulted in incorrect product and process design decisions.

A second important question is the extent to which cost system designers should allow motivational effects to dominate the need for accurate product costs. First, the cost system is frequently the only source of information on product costs. Second, designing a cost system solely to motivate behavior can, as it did at Zytec, lead to acceptance problems because reported product costs lack credibility. Finally, internal decisions can be influenced by other means such as performance measurement systems or by legislating policy. At Zytec, for example, a performance measurement system was used to reduce cycle time. At PID, designer behavior was modified by setting policies for new designs, such as a limited set of parts, limiting the number of subassemblies, and eliminating the need for repositioning the product on the bench during assembly.

We recommend that a firm, prior to designing an activity-based cost system, examine the main source of potential benefit from installing the system. If the main benefit is external, the system should be designed with sufficient cost drivers to provide reasonably accurate product costs. If the main benefit is improved internal decisions, the system should be designed to clearly communicate the link, via the cost drivers, between these internal decisions and cost. The designers of internally focused systems must be careful, however, not to sacrifice too much accuracy. This is particularly true in firms where cost information is used for external as well as internal purposes. Accuracy should only be decreased when the benefits of sending a clearer message outweigh the costs of reporting less accurate cost information.

REFERENCES

Cooper, R., and R. S. Kaplan, 1988a. "How Cost Accounting Distorts Product Cost." *Management Accounting* (Vol LXIX No. 10 April 1988), pp. 20-27.

Cooper, R., and R. S. Kaplan, 1988b. "Measure Costs Right: Make the Right Decisions." *Harvard Business Review* (September-October 1988), pp. 96-103.

Cooper, R., 1988a. "The Rise of Activity-Based Costing—Part Two: When Do I Need an Activity-Based Cost System?" *Journal of Cost Management for the Manufacturing Industry* (Vol 2, No. 3 Fall 1988), pp. 41-50.

Cooper, R., 1988b. "The Rise of Activity-Based Costing—Part Three: How Many Cost Drivers Do You Need, and How Do You Select Them?" *Journal of Cost Management for the Manufacturing Industry* (Vol 2, No. 4 Winter 1989), pp. 34-46.

Cooper, R., 1988c. "The Rise of Activity-Based Costing—Part Four: What Do Activity-Based Cost Systems Look Like?" *Journal of Cost Management for the Manufacturing Industry* (forthcoming).

Hayes, R. H., and S. C. Wheelwright. "Linking Manufacturing Process and Product Life Cycles." *Harvard Business Review* (January-February 1979), pp. 133-140.

WORLD-CLASS MANUFACTURING: PERFORMANCE MEASUREMENT

Robert W. Hall
Indiana University

The new manufacturing is frequently described as a mix of just-in-time manufacturing, Total Quality Control and Employee Involvement, or in words that mean the equivalent. One of the most common barriers to redirecting a company toward the new manufacturing is often said to be the performance measurement system. The implication is that the performance measurement system is somehow immutable, but that is not so. In human organizations we measure what we believe to be important, and many rational-sounding discussions about conflicts in performance measurements are really a veneer covering a lack of common understanding or agreement about the real operational goals of the company.

We need strategic performance measurement systems — strategic in the sense that we define the goals toward which the company should work. Then goals can be stated in a way that progress toward them can be measured. Even when measuring costs, we need to begin by reflecting on the purpose for which a cost figure will be used; whether for strictly internal comparisons of resource use, for external comparisons, or as a basis for pricing decisions, and so forth.

THE GOALS OF WORLD-CLASS MANUFACTURING

Goals begin with customer satisfaction. Without that, excellence in internal operations mean little. That in itself covers a broad scope which classical markets try to span with the four "P's": Product, Price, Place and Promotion. Translated to slightly more operational terms, that is: Product, price, quality, delivery, service and responsiveness. In still more operational terms, the goal is to give customers what they want when they want it without waste. Simple as that sounds, that approach to the customers demands the utmost development of company operations.

Rehashing goals seems wasteful to many who want to get on with the activity of JIT and TQC, but many of the corrupted versions of both JIT and TQC start from a narrow interpretation of their goals. For example, the belief that the goal of JIT is only to reduce inventory investment leads to pushing inventory out to the suppliers. Total system inventory and lead time may be little changed. Sometimes TQC is interpreted as merely measuring the defects by a statistic. Overall goals need to be total system improvement goals, and measurements need to be total system measurements.

Figure 1 shows several well-accepted goals of the new manufacturing, accompanied by some measures of overall performance toward each goal. The sixth goal, continuous improvement, is not a specific performance goal, but a description of the kind of process which is desired of the people within the company. Continuously developing people means continually setting new objectives and striving to meet them. If this is done, performance trends should show improvement. If a disturbance interrupts improvement, the matter is corrected and performance resumes improvement again.

For this reason the examination of trends is very important. When performance begins to plateau, the improvement process needs another stimulus.

Companies cannot pay attention to the improvement of everything at once. They must typically improve in stages. For example, they concentrate on improving quality while also improving on other measures, or at least not regressing on other measures. For instance, a company may streamline the flow of material through a plant, then try to hold leadtimes constant while working on standardization of processes to improve quality. There is a relationship between consistency in leadtimes and consistency in methods, and consistent methods promote consistent quality.

Companies need an improvement program which typically moves through several stages, and performance measures should help guide their progress through it. Since the improvement program should undergird the competitive position of the company, a reasonable name for this philosophy toward performance measures is "strategic performance measures." The operating performance measures derived from this are seldom the same as those which are intended to show whether the company is making a good return for the owners.

Many financial measures are really intended to show what operations are contributing toward profitability. In many ways the accounting system is derived from the need to show the status of the company to owners. The resulting performance measures thus emphasize cost control or cost minimization, where cost is viewed as the number derived from the cost system itself. (In viewing reduction of resource use as a goal, cost is a measure of resource use and so are various ratios, but waste denotes the ineffective use of resources in useless activities itself. Sometimes waste can be seen merely by eye — direct sight. No abstract measure is required to spot it.)

In addition, cost minimization is so time-honored a tradition of performance measurement in manufacturing that measures must be translated to cost in order to become legitimate. Two such measures are cost of quality and cost of inventory. Both are based on relatively murky parameters or redistribution of cost numbers by difficult-to-define quality loss ratios. If a management has faith that defect rates should be reduced, there is little need for them to be further stimulated by a cost of quality number. Removal of waste should show up in lower cost numbers sooner or later, if real costs are consistently measured.

NEW MEASURES OF PERFORMANCE

One of the most difficult to understand measures is the concept of leadtime as an indicator of waste. Inventory measured in days on hand is a surrogate for leadtime. A value added ratio is more revealing of the relationship between leadtime and waste, but is also more difficult to comprehend. One approximation of a value added ratio is to contrast the work time for one piece (not a large lot) against the total elapsed time the part goes through the total process. Strictly interpreted, the ratio is very small, seldom over one percent for a single piece part if an approximation can be made at all.

Another approximation is derived by flow-charting the part all the way through the process. Then note all the steps in the flow that truly add value versus the total number of steps identified. The number of true value-adding operations may be as low as one or two of a hundred. The rest are candidates for elimination. The method is a good way to identify waste also, provided time is not wasted doing excessive flow-charting.

A more extensive version of this is possible if cycle times can be defined as the time between recurring events, and an automotive assembly line is considered as an example. Take a line running at 60 cars per hour. That is a one minute cycle time between completions of cars. Suppose there are 1200 cars on line. That is 20 hours of cars, but it is more useful to think of it as 1200 potential one minute cycles of work. Of those potential cycles, how many are actually used? That value-added ratio is still an overestimate because not all the time is actually used within each work cycle at each station. Much of the time is wasted with material handling or other activity. Only a fraction of the time is spent in actual attachment-and-conversion build-up of the vehicle.

We are accustomed to studying productivity of machines and productivity of people. We are not so accustomed to thinking about the productivity of the time during which we have access to work, but do not always do it. Yet that is a highly productive way to look at it. In a recent study which compared a Japanese and an American auto assembly plant, one difference was that the Japanese workforce concentrated on minimizing the amount of time a car merely sat on line. They pursued root causes into quality problems and material positioning and worker flexibility. Value-added ratios could not be obtained for comparison, but the American plant had 24 hours of cars on line; the Japanese plant, 10.8 hours. Partly as a result of concentrating on the activity associated with the cars themselves, the Japanese plant enjoyed a direct labor productivity advantage at least three times greater than that of the American counterpart. A major difference in thinking: The American dictum was to not let the line stop, thus turning off the revenue spigot. The Japanese idea was to never let any car body merely sit *anywhere* in the process without doing something to it. The Americans pressed for completed results; the Japanese for a high value-added process. Even the Japanese practice of Jidoka, stopping the line when in trouble, had the purpose of quickly restoring the process to a high value-adding condition.

A good value-added measure is one that suggests ways to further value-adding activity, that is, ways to increase the quality and suitability of *all* units in the process as quickly as possible. They would be more total system measures than machine utilizations or efficiencies. Those measures assume that operations are independent, and they are not. Manufacturing consists of a multitude of interrelated activities.

The measures do not need to trigger improvement ideas directly, but they should cause brainstorming to find ways of decreasing waste and leadtime. Westinghouse uses a time-cost profile they call OPTIM to trigger this kind of activity. General Electric uses time-scaled flow charts.

Time-line measures are growing in popularity, but slowly. Even most of the users of them still have trouble fully appreciating *why* decreasing the time to process material (or anything else) will create improvement on many other measures.

Quality Measurement

Quality is defined in different ways. One of the most common is that quality is conformance to specification, but that assumes that the specifications were developed from a quality process. In the broadest sense, quality is pleasing the customer, so even measures of customer satisfaction come under the heading of quality, and quality performance consists of coming ever closer to the marks needed to please the customer, whether the customer is explicitly aware of the technical requirements to meet their needs or not.

Quality measures can be divided into three (or more) groups: External quality as seen by the customers or others outside the company; internal quality of the operations and processes inside the organization; and the nature of the company's quality improvement processes themselves. Examples of external quality measures are direct survey returns from customer, warranty rates, reliability, service call effectiveness, and so on. Internal quality measures are overall yields, defect rates, re-work rates, inspection ratios, process capabilities and so forth.

These measures have become well-known. Just as important is to measure the nature of the quality improvement process itself. Are quality processes taught? Is group and organizational problem solving actually a part of daily life? Do individual employees understand quality tools, and do they practice a form of quality improvement in daily work? How greatly are quality practices deployed throughout the company's employees and representatives, and in how many operations? These are all good questions, but how to measure the "answers" is part of making the results positive.

Individual measurements of the quality improvement process are not so hard. What percentage of employees have had training of various types? How many "story-boards" of improvement projects are underway? What percentage of employees are participating in quality improvement activities? What is the trend in awards and recognition for quality improvement successes? Perhaps most important is whether the organization is meeting specific milestone accomplishments in the development of quality.

Summarizing the status of the overall quality improvement process into one or two indicators is more difficult. A contribution to this has been made by the measurement process for the Malcolm Baldrige National Quality Award. Each criterion judged for the award is scaled from zero to 100 percent in 10 percent increments. A zero percent score indicates that the company's effort on that criterion is only talk and no action. A 50 percent score indicates that a good methodology is in place, some activity is occurring and results are about as expected for the level of effort thus far expended. For example, if a machining company began SPC three years ago and only 10 percent of their processes had C_{pk}'s above 1.33 then, but half of them have achieved at least that mark today, then significant progress has been made. It does not represent great progress narrowing variances to a targeted value, but it is good progress on an admittedly difficult task of upgrading much tooling and equipment.

A 100 percent score on this same process capability criterion would be truly outstanding. The company would have 100 percent of all processes at a C_{pk} of 1.33 or more and be working to narrow variances still further. There would be no question that this objective was desired so long as the expense of reaching were not great. That is the decision making would be more on quality than on lowest measured cost for all alternatives. Such thinking would be deployed everywhere in the company, and it would have been sustained long enough to be a way of life with employees and an exemplary practice for other companies.

Judging process capabilities is only one of many, many criteria that could be evaluated in assessing overall quality of product, process and service. In effect the Baldrige evaluation makes an overall weighted evaluation of all criteria applicable to the company (as much as can be determined by the evaluators using a carefully developed set of criteria). One of the major uses of the Baldrige Award criteria and scoring system in the future may be for companies to give themselves an evaluation using a set of criteria which is as close as the United States has ever had to a consensus on the definition of quality.

By this standard of measuring, an overall 50 percent company is one in which there are pockets of excellence, many of which are good enough to be the subject of seminars, cases and workshops. There are today many such companies in the United States. Within 5 or 10 years, a 50 percent rating may be the minimum requirement to hold position in international competition. The competence of practice is sharply increasing everywhere, not just in Japan. The companies which will be taking business from others should within 10 years be closer to 100 percent companies.

Leadtime Measurement

Leadtimes are seldom measured directly. They are approximated because, for instance, it takes considerable attention to actually mark a piece of material as it enters the plant and check it as it leaves, buried inside an assembled product or formed into a different shape and painted. However, it can be done, and occasionally it is done. Most plant throughput times are approximated from WIP inventory levels. Order leadtimes are estimable from the dating of the orders provided it is considered important to measure this leadtime. The question is what leadtimes are important and why?

The objective of decreasing leadtimes is to stimulate the removal of waste from the processes first. Later, an objective is to make actual changes as quickly as possible without waste — without extra inventory or expensive means.

For example, why reduce supplier leadtimes? Because doing so forces both the customer and supplier companies to examine the quality, inspections, checks, reviews and other non-value-adding activities that create the leadtime. Beyond that, the purpose is to enable changes to be made as quickly as possible. That is, a true supplier leadtime is not the leadtime to execute according to plan, but the test of real performance is being able to increase or decrease total volume or change the mix of parts shipped quickly and without a stocking plan.

Ability to change quickly is the basis of flexible response to a changing market — without waste, generally in the form of inventory built to a bad forecast. Checking for the time necessary before a change in volume or mix can come from production is an exercise more frustrating than just expressing inventory in a days-on-hand figure, but it gives a better notion of the true condition of the manufacturing company or its suppliers. Unless a change really needs to be executed, estimating time-to-change is a what-if exercise that does not need to be repeated often, perhaps once or twice a year. However, the estimates thus taken are indicators of an operating capability that is worth paying attention to — and attempting to develop.

Flexibility is a largely undiscussed aspect of just-in-time manufacturing that deserves more attention. It comes to the fore as soon as one begins to investigate the causes of long order leadtimes in make-to-order businesses. The causes will typically involve the inability to locate, shift, repair or even modify equipment and tooling, either within the customer plant or at supplier plants. Therefore, some very important leadtimes to measure are:

- Tooling turnaround time.
- Equipment repair time.
- Time to change layout.
- Engineering change time.
- Tooling design time and tooling build time.

Measures of People Development

One of the needs of just-in-time production is for multifunctional personnel. That begins with a skills inventory. Few companies have a complete skills record on all their people, including direct workers, that may include skills acquired in previous jobs or in outside activities. Some skills are of little value, but some are also surprising.

Many Japanese companies display in each department a roster of all employees there and the skills or tasks for which they are qualified. The practice is now occasionally seen in the United States. Many companies have found it necessary to extend the cross-training of people in breadth and depth when moving from single work-station to cell work, or when starting preventive maintenance. Just as important when stimulating group problem solving is recording the development of people in group activities. Leadership or facilitation training certificates are important.

Recognition is also very important. Most of the companies with strong employee involvement programs have a proliferation of employee recognition materials in evidence. These require a great deal of imagination to keep fresh and at the same time avoid degenerating into displays seen as silly. The contributions of people are every bit as important to keep as part of performance measurement as some of the production numbers.

Resource Use

The most common measures of resource use are from cost accounting, but there are others: Productivity ratios, space utilization, machine utilization and materials consumption rates to name a few. An example of materials consumption important in some cases is rate of use of cooling water.

All these measurements have their uses and caveats about use. For instance, when comparing two operations on the basis of labor productivity, one must be careful that the same segment of the workforce is compared in both instances, that the work content in each facility accomplishes the same conversion effects — similar degrees of vertical integration — and so forth.

However, much of the controversy in recent years has come from the use and misuse of cost figures. No costing system perfectly models the use of resources in production. It is unreasonable to expect such a model. The question is whether cost models are adequate, and whether they are properly used.

The issues begin with the purpose of cost figures. Most systems are derived from the financial reporting system used to evaluate whether the company is making a profit. The systems seem to accomplish that objective, provided the accounting cycles are long enough to give a picture of true operational change between the times of measurement. That is, given the problems of allocation and of accrual approximations, looking at monthly changes too closely may only be studying measurement "chatter." In quality control terms, the changes in the process can be masked by the errors of measurement.

However, existing cost systems may not give very good guidance on the actions to take to be competitive. The problems of cost systems have been well-aired recently. From the viewpoint of someone attempting a total operations improvement process, several of the weaknesses stand out:

1. Most cost systems are part of product costing systems in which large overhead costs are allocated on the basis of direct labor. Direct labor is not a very good indicator of the distribution of resource use on other efforts associated with the company's mix of products.
2. Cost systems assume that various operations are independent — that one has little affect on another. That is not true, and as operations are meshed more closely together, it becomes less true. Changes in one operation affect the workings of other operations. (One way to reflect on this is to make a detailed flow diagram of the many steps in the flow of a part through a manufacturing process. Then ask how many steps are directly captured by the cost system and how many fall into the broad categories called overhead.

Overhead costs are high, but so would be the cost to capture extensive resource use detail, and data recording is in itself an activity that consumes time and thus slows the flow of work.)

3. Many systems are variance measurement systems. The variances are measured as deviations from costs projected as derived from the plant budget. There are two problems with this approach. First, the results of variance analysis are sometimes confused because one does not know if the "problem" is a result of poor operation or poor budget targets. This is compounded if variance figures are calculated long after the action is done. Then explanations of variance become even more of an excuse report. Second, many variances are detailed, and problems affecting one cost variance affect several, but the interaction is not always clear, especially if the operations are in two separate parts of an organization. The result is the classic "finger point" explanation.

4. A compilation of costs useful for pricing, external reporting and making various kinds of operating decisions is not always the same set of costs. An "official" cost system used for multiple purposes may lead to numerous "Catch-22" situations.

Given numerous problems, what should be done? Changes in cost systems tend to be expensive. Perhaps the first action is about the same as that taken with any system which appears to be outliving its usefulness: Abandon the parts of it which cause real trouble, and wait to see what kind of process changes take place before expending enormous effort developing a new system. A second action is to not make operating decisions strictly on the basis of small cost differences between alternatives. Have faith enough to straighten flows, cut defect rates and move in a strategically useful direction even if cost figures say there is little or no payback. Over time, if the total operation uses fewer resources, the result should show as a decreased overall actual cost no matter what kind of cost system is used. Cost is a veto of an operating improvement only if judgement says the cost is ridiculous — not a close call. Generally common sense should warn that decreasing defect rates or decreasing leadtimes should not be done with a big expense. If so one must question if waste is really being removed from the system.

Ideally, the cost system of a JIT company should promote that company's effort. Such a cost system should be used to help determine if the company is moving in the direction it wants to go.

Activity-based costing (ABC) shows promise of helping with these problems, but there is no panacea — no magic technique that can replace dedication and management effort. That is, there is no accounting formula that can reduce a JIT or TQC process to a management-by-the-numbers methodology where the numbers are only costs — measures of resource use.

For example, suppose a company sees that it can reduce billing errors by having sales people ask for purchase order numbers to be repeated and verified. That added step will increase each customer contact time by a few seconds. It may then be construed as reducing the available time to advance-sell or cross-sell the customer, so perhaps the company is better off if the errors are simply caught and worked out later. In this case, as with so many, the value of both making a change and not making a change is nebulous, and especially when estimated in advance. There is simply a certain amount of a priori faith required that cutting *total* leadtimes and *total* error rates will have great payback, however vaguely that payback may be measured.

After a year or two of honest effort, the real payback should be evident in a reduction in total cost, or a reduction in cost of goods sold. At that point, it may be impossible to attribute the cost improvement to any one change. However, one should be able to see benefit from a change in strategic direction — on all the measures previously mentioned and on cost.

Figure 1. Overall Goals and Total Performance Measures

Goal	Examples of Total System Measure
1. Eliminate Waste	1. Ratio of total no. of value-added operations to the total operations on a supplier-to-customer material flow chart.
2. Improve Quality	2. a. External: Warranty return rates. b. Internal: Trend in overall process yield.
3. Reduce *All* Leadtimes	3. a. Total system production leadtime, supplier to customer delivery. b. Customer leadtime: Order booking to product delivered and functioning.
4. Reduce Resource Use	4. Total actual unit cost.
5. Develop People	5. a. Increase in value-added per person. b. Suggestions implemented per employee.
6. Continuous Improvement	6. Check whether *trends* of measures such as those above continue to improve.